MULTICARRIER MODULATION
WITH LOW PAR
Applications to DSL and Wireless

**THE KLUWER INTERNATIONAL SERIES
IN ENGINEERING AND COMPUTER SCIENCE**

MULTICARRIER MODULATION WITH LOW PAR
Applications to DSL and Wireless

by

Jose Tellado
Stanford University

KLUWER ACADEMIC PUBLISHERS
Boston / Dordrecht / London

Distributors for North, Central and South America:
Kluwer Academic Publishers
101 Philip Drive
Assinippi Park
Norwell, Massachusetts 02061 USA
Telephone (781) 871-6600
Fax (781) 681-9045
E-Mail <kluwer@wkap.com>

Distributors for all other countries:
Kluwer Academic Publishers Group
Distribution Centre
Post Office Box 322
3300 AH Dordrecht, THE NETHERLANDS
Telephone 31 78 6392 392
Fax 31 78 6546 474
E-Mail <services@wkap.nl>

   Electronic Services <http://www.wkap.nl>

Library of Congress Cataloging-in-Publication Data

Tellado, Jose.
　Multicarrier modulation with low par : applications to DSL and wireless / by Jose Tellado.
　　p.cm.
　Includes bibliographical references and index.

　　1. Digital communications. 2. Wireless communication systems. 3. Carrier waves. 4.
　Modulation (Electronics) I. Title.

　TK5103.7 .T463 2000
　621.382—dc21

00-062494

ISBN 978-1-4419-5009-3　　　　　e-ISBN 978-0-306-47039-4

Printed on acid-free paper.

Printed in the United States of America.

To Louise

Contents

List of Figures

Foreword

With the release of this book, Dr. Jose Tellado, provides a visionary view of the solutions for the reduction of the notorious peaks that have previously plagued multicarrier transmission systems. The author of this foreword has been continuously amazed at the progress Tellado made in eliminating from concern this purported achilles' heel of multicarrier transmission. Tellado goes so far with peak reduction as to show that the peak-to-average power ratio of a well-designed multicarrier system is actually less than what it is in sophisticated single-carrier designs. Described in this book are several very tangible and realizable methods for corrrect, simple design of the multicarrier system. We stand at the threshold of a future multicarrier transmission world, predicted by Shannon and materialized by a series of young talents like Tellado, who is without a doubt, the preeminent authority in the world on peak-to-average power ratio reduction in multicarrier transmission.

Tellado's work has impacted international DSL standards, particularly G.992 ITU standards (G.lite and G.dmt) and the emerging G.vdsl, and will continue to impact future standards, probably extending into wireless transmission and cable-modem transmission as well. The ideas and methods are novel and creative, and begin with an introduction to one of the easiest methods to implement − tone reservation. Tellado's tone reservation methods create a mechanism for using a small fraction of bandwidth to synthesize signals that are of opposite polarity and shape to a peak in a multicarrier signal. Subtraction of such a peak then follows easily without effect upon the data-carrying capability of the multicarrier signal. Enormous reduction of peak signals is possible with tone reservation methods at low implementation cost.

Tellado proceeds into systems that have tiny data-rate loss without reservation of tones that allow constellation extension by dummy points so that signals with lower peaks can be constructed. While more complicated than tone reservation, these newer tone-injection systems allow essentially zero rate loss in transmission to reduce peak energy very significantly. These injection methods require no coordination between transmitter and receiver, unlike the tone-reservation methods where the receiver needs to agree with the transmitter on the tones to be used for peak-reduction in a multicarrier transmission system.

Finally, Tellado proceeds to nonlinear receiver structures that allow the mitigation of nonlinear clipping in a transmitted signal. That is peaks are simply allowed to occur in transmit-signal generation, but clipped by analog electronics. The consequent distortion is recognized and eliminated by the receiver.

The net peak-to-average reduction possible with the methods of Tellado makes multicarrier transmission extremely attractive even in situations with several amplifier constraints. The understanding and analysis of peak issues in this book is outstanding, a credit to Tellado and an enormous aid to the reader. This is a must-read book for the serious designer of multicarrier transmission systems, and the beginning material from a young career that will proliferate creative solutions for difficult transmission problems.

John M. Cioffi
Stanford University

Preface

T HE DATA RATE AND RELIABILITY REQUIRED to support the new information age has increased the demand for high speed communication systems. Multicarrier modulation has recently gained great popularity due to its robustness in mitigating various impairments in such systems.

A major drawback of multicarrier signals is their high Peak to Average Power Ratio (PAR). Since most practical transmission systems are peak-power limited, the average transmit power must be reduced for linear operation over the full dynamic range, which degrades the received signal power.

This book formulates the PAR problem for multicarrier modulation and proposes three new methods for PAR reduction. The first two structures prevent distortions by reducing the PAR on the discrete-time signal prior to any nonlinear device such as a DAC or a power amplifier. The third structure corrects for transmitter nonlinear distortion at the receiver when the nonlinear function is known. Most methods reduce the PAR of the discrete-time signal although the PAR of the continuous-time signal is of more interest in practice. We derive new absolute and statistical bounds for the continuous-time PAR based on discrete-time samples.

The first distortionless structure, called Tone Reservation, reserves a small set of subcarriers for reducing the PAR. For this method, the exact solution and several low complexity suboptimal algorithms are presented.

The second distortionless structure is called Tone Injection. For this method, the constellation size is increased to allow multiple symbol representations for each information sequence. PAR reduction is achieved by selecting the appropriate symbol mappings. The exact solution to this PAR minimization problem has non-polynomial complexity. Bounds on maximum PAR reduction are derived and efficient algorithms that achieve near optimal performance are proposed. Similar to the Tone Reservation method, most of the complexity is introduced at the transmitter. The additional complexity at the receiver is a simple modulo operation on the demodulated complex vectors.

The third structure reduces the PAR by applying a saturating nonlinearity at the transmitter and correcting for nonlinear distortion at the receiver. This simplifies the transmitter at the expense of adding complexity at the receiver. Mutual Information expressions are derived for multicarrier transmission in the presence of nonlinear distortion. The optimum maximum likelihood receiver and an efficient demodulator based on the maximum likelihood receiver are also described.

JOSE TELLADO

Acknowledgments

T HE WORK IN THIS BOOK was done over the last two years of my graduate studies at Stanford, and I have many people to thank for helping to make my Stanford experience both enriching and rewarding.

First and foremost, I wish to thank my adviser, Prof. John Cioffi. Throughout my years at Stanford, John has been a constant source of research ideas, encouragement and guidance. His confidence in me, his inspiring words in times of need and his unique blend of academic and entrepreneurial attitude have created an ideal environment for this research.

I also wish to thank my associate adviser, Prof. Donald Cox. His breadth of knowledge and perceptiveness have instilled in me great interests in wireless communications. The quality of this book has also improved due to his careful review and many good suggestions.

I am grateful for the financial support from Apple Computers, Accel Partners and Intel. Without their generous donations, this research would not have been realized.

It was a privilege and a pleasure to be one of Cioffi's "kids". Given the size of the group, it's amazing how well we all get along. Members of this extended family whose friendship and support I have enjoyed are, in alphabetical order, Carlos Aldana, Phil Bednarz, Kok-Wui Cheong, Won-Joon Choi, Pete Chow, John Fan, Minnie Ho, Krista Jacobsen, Joonsuk Kim, Joe Lauer, Acha Leke, Susan Lin, Ardavan Maleki, Rohit Negi, Greg Raleigh, Wonjong Rhee, Atul Salvekar, Rick Wesel and Nick (Zi-Ning) Wu. I am also grateful to Joice DeBolt for her invaluable assistance with all administrative work.

Outside of John's research group, I have been very fortunate to engage in significant discussions with a group of very talented people. They are Suhas Diggavi, Bijit Halder, Miguel Lobo and Jorge Campello. Their insights have helped my research, but more importantly, I cherish their friendship. My thanks also go to Carlos Mosquera, Josep Maria Añon, Jaime Cham, Gabriel Perigault, Angel Lozano and Daniel Ilar who have remained great friends since the early days at Stanford.

Through it all, I owe my greatest debt to my family — my parents, my brother Tony and my sister Monica. I am most grateful for their love, encouragement, and support throughout my life.

To Louise, who has given me happiness and unwavering support in my life and work, thank you. Your love is the most rewarding experience of my graduate years.

As I embark on new challenges in life, I know I will always reflect on my time at Stanford with great nostalgia.

Jose Tellado

About the Author

Jose Tellado received the M.S. and Ph.D. degrees in Electrical Engineering, with an emphasis on Communication Theory, from Stanford University in 1994 and 1999, respectively. From 1994 to 1999, he was a Research Assistant at the Information Systems Laboratory at Stanford University where he proposed new advanced methods for Multicarrier Modulation (DMT/OFDM) in the areas of Peak to Average power Ratio reduction, and Multiuser Transmit Optimization. While at Stanford, he held consulting jobs at Apple Computers Advanced Technology Wireless Group, where he researched and modeled the GMSK-based Physical Layer called HIPERLAN, and at Globalstar, where he analyzed and modeled their IS-95 based Low-Orbit Satellite cellular system.

In early 1999, he joined Gigabit Wireless, then a very early-stage broadband wireless access start-up. There, he is the chief designer of the Physical Layer, and has actively participated in the design of the wireless system architecture, the Medium Access Control Layer and the RF specification.

He has over 12 IEEE publications, 10 xDSL standard contributions and 6 pending patents.

Chapter 1

INTRODUCTION

T HE DEMAND FOR HIGH DATA RATE SERVICES has been increasing very rapidly and there is no slowdown in sight. Almost every existing physical medium capable of supporting broadband data transmission to our homes, offices and schools, has been or will be used in the future. This includes both wired (Digital Subscriber Lines, Cable modems, Power Lines) and wireless media. Often, these services require very reliable data transmission over very harsh environments. Most of these transmission systems experience many degradations, such as large attenuation, noise, multi-path, interference, time variation, nonlinearities, and must meet many constraints, such as finite transmit power and most importantly *finite cost*. One physical-layer technique that has recently gained much popularity due to its robustness in dealing with these impairments is multicarrier modulation.

Unfortunately, one particular problem with multicarrier signals, that is often cited as the major drawback of multicarrier transmission, is its large envelope fluctuation, which is usually quantified by the parameter called Peak-to-Average Ratio (PAR). Since most practical transmission systems are peak-power limited, designing the system to operate in a perfectly linear region often implies operating at power levels well below the maximum power available. In practice, to avoid operating the amplifiers with extremely large back-offs, occasional saturation of the power amplifiers or clipping in the digital-to-analog-converters must be allowed. This additional nonlinear distortion creates inter-modulation distortion that increases the bit error rate in standard linear receivers, and also causes spectral widening of the transmit signal that increases adjacent channel interference to other users. This book derives a number of relevant results regarding PAR analysis and designs a number of efficient, low complexity PAR reduction algorithms.

While most derivations and results are described for multicarrier modulation based on the discrete-time or continuous-time Fourier basis, many of these results can be extended to other multicarrier modulation methods, such as Vector Coding, Discrete Wavelet MultiTone, etc. Moreover, some results can be applied to non-multicarrier modulations, such as single carrier modulation or Direct-Sequence Spread Spectrum to name a few. For the single-carrier case, the transmit basis is generated from the equivalent transmit pulse shaping function. If this transmit pulse shaping function has considerable energy over several symbols, the three main PAR reduction ideas described in this book can be applied, although the algorithms described must be modified. Similarly, many of our PAR reduction ideas can be utilized whenever multiple Direct-Sequence Spread Spectrum signals or multiple single carrier signals are combined.

1.   OUTLINE OF BOOK

Chapter 2 introduces multicarrier modulation and more specifically the two most commonly used orthogonal multicarrier modulation types, Discrete MultiTone (DMT) and Orthogonal Frequency Division Multiplexing (OFDM).

Chapter 3 describes real and complex multicarrier signals using continuous-time and discrete-time formulations and formally introduces the concept of PAR. This chapter also derives bounds on the PAR of the continuous-time symbols given the discrete-time samples. Since the PAR reduction methods described in Chapters 4 and 5 are simplified when operating on discrete-time symbols, these bounds can be used to predict the continuous-time PAR. Also, several common nonlinear models are presented and their effect on multicarrier demodulation is evaluated.

Chapter 4 presents the first new distortionless PAR reduction structure, called Tone Reservation. The exact solution to this PAR minimization problem, as well as efficient suboptimal solutions are described followed by a number of simulation results for practical multicarrier systems.

Chapter 5 describes a second distortionless PAR reduction structure, called Tone Injection. Since the exact solution to this PAR minimization problem has non-polynomial complexity, limits on maximum PAR reduction are derived. Efficient suboptimal solutions that achieve close to optimal performance are proposed.

Chapter 6 evaluates the performance of multicarrier transmission in the presence of nonlinear distortion. Theoretical Mutual Information limits are derived. The optimal maximum likelihood receiver and an efficient demodulator based on the maximum likelihood receiver are also described.

Chapter 2

MULTICARRIER MODULATION

A CHIEVING DATA RATES that approach capacity over noisy linear channels with memory, requires sophisticated transmission schemes that combine coding and shaping with modulation and equalization. While it is known that a single-carrier system employing a minimum-mean-square error decision-feedback equalizer can be, in some cases, theoretically optimum [Cioffi et al., 1995], the implementation of this structure in practice is difficult. In particular, the required lengths of the transmit pulse shaping filters and the receiver equalizers can be long and the symbol-rate must be optimized for each channel. An alternative scheme that is more suitable for a variety of high-speed applications on difficult channels, is the use of multicarrier modulation, which is also optimal for the infinite-length case. The term multicarrier modulation includes a number of transmission schemes whose main characteristic is the decomposition of any wideband channel into a set of independent narrowband channels. Within this family, the most commonly used schemes are Discrete MultiTone (DMT) and Orthogonal Frequency Division Multiplexing (OFDM), which are based on the Discrete Fourier Transform, resulting in an implementable, high performance structure.

The main focus of this chapter is to introduce some of the basic ideas of multicarrier modulation and provide the necessary background for understanding the subsequent chapters. More detailed descriptions can be found in listed references.

1.   MULTICARRIER MODULATION

All multicarrier modulation techniques are based on the concept of *channel partitioning*. Channel Partitioning methods divide a wideband, spectrally shaped transmission channel into a set of parallel and ideally independent narrowband subchannels. Although channel partitioning is introduced here, the reader is referred to [Cioffi, 2000b, Chapter 10] for a more in-depth tutorial description of channel partitioning for both the continuous-time case and the discrete-time case. For the continuous-time case, the optimum channel-partitioning basis functions are the set of orthonormal eigen-functions of the channel autocorrelation function. These eigen-functions are generally difficult to compute for finite symbol periods and are not used in practical applications. Instead, we will focus on discrete-time channel partitioning, which partitions a discrete-time description of the channel. For this case, it is assumed that the combined effect of transmit filters, transmission channel and received filters can be approximated by a finite impulse response (FIR) filter. Such a description is not exact, but can be a close description as long as a sufficient number of samples for the input and output of the channel is included and good timing and frequency synchronization is achieved [Moose, 1994, Schmidl and Cox, 1997].

Calling $\mathbf{h} = [h_0 \cdots h_\nu]$ the baseband complex discrete-time representation of the channel impulse response, the block of output samples $\mathbf{y} = [y_0 \cdots y_{N-1}]$ can be expressed as a function of the input samples $\mathbf{x} = [x_{-\nu} \cdots x_{-1} \; x_0 \cdots x_{N-1}]$ and the channel noise $\mathbf{n} = [n_0 \cdots n_{N-1}]$ using the following standard vector representation,

$$
\begin{bmatrix} y_{N-1} \\ \vdots \\ y_0 \end{bmatrix} = \begin{bmatrix} h_0 & h_1 & \ldots & h_\nu & 0 & 0 & \ldots & 0 \\ 0 & h_0 & h_1 & \ldots & h_\nu & 0 & \ldots & 0 \\ \vdots & \ddots & \ddots & \ddots & \ddots & \ddots & & \vdots \\ 0 & \ldots & 0 & h_0 & h_1 & \ldots & & h_\nu \end{bmatrix} \begin{bmatrix} x_{N-1} \\ \vdots \\ x_1 \\ x_0 \\ \vdots \\ x_{-\nu} \end{bmatrix} + \begin{bmatrix} n_{N-1} \\ \vdots \\ n_0 \end{bmatrix} \quad (2.1)
$$

This can be written more compactly as

$$
\mathbf{y} = H\mathbf{x} + \mathbf{n} \quad (2.2)
$$

Equation (2.2) represents the input-output relationship for a single block of samples. In a practical communication system, we may transmit in a continuous manner and (2.2) becomes $\mathbf{y}^m = H\mathbf{x}^m + \mathbf{n}^m$ where the index m represents the m-*th* block.

To send information over a finite-length multicarrier system, the data must be partitioned into blocks of bits, and each block must be mapped into a vector of complex symbols $\mathbf{X}^m = [X_0^m \cdots X_{N-1}^m]$. The modulated waveform $\mathbf{x}^m$ is

$$\mathbf{x}^m = \sum_{k=0}^{N-1} \mathbf{m}_k X_k^m = M\mathbf{X}^m \qquad (2.3)$$

where $\{\mathbf{m}_k, k = 0, \ldots, N-1\}$ denote the set of transmit basis vectors and M is the matrix constructed with the transmit basis vectors as its columns. At the receiver, the received vector $\mathbf{y}^m$ is demodulated by computing

$$\mathbf{Y}^m = \begin{bmatrix} \mathbf{f}_{N-1}^* \mathbf{y}^m \\ \vdots \\ \mathbf{f}_0^* \mathbf{y}^m \end{bmatrix} = F^* \mathbf{y}^m \qquad (2.4)$$

where $\{\mathbf{f}_k^*, k = 0, \ldots, N-1\}$ denote the set of receive basis vectors and F^* is the matrix constructed with the receive basis vectors as its rows. The notation A^* indicates the hermitian of matrix A, which represents the conjugate of the transpose of A. The overall input-output relationship is given by,

$$\mathbf{Y}^m = F^* H M \mathbf{X}^m + F^* \mathbf{n}^m \qquad (2.5)$$

Different choices for F and M are possible, leading to a number of possible multicarrier structures. For the simple case where the channel is memoryless, $\mathbf{h} = [h_0 \ 0 \cdots 0]$, choosing M to be an orthogonal matrix i.e. $M^{-1} = M^*$, and setting $F = M$ result in perfect channel partitioning. That is, the output vector components can be expressed as a function of only one input component and (2.5) can be simplified to N scalar equations:

$$Y_k^m = h_0 X_k^m + N_k^m, \quad k = 0, \ldots, N-1 \qquad (2.6)$$

where the noise samples N_k^m are independent identically distributed (i.i.d.) Gaussian samples if the time domain noise samples n_k are i.i.d. Gaussian samples. For non-trivial channels, a variety of multicarrier structures have been proposed, each with a different choice of basis vectors. Sections 2. and 3. describe two asymptotically optimal multicarrier structures called Vector Coding [Aslanis, 1989, Kasturia, 1989, Kasturia et al., 1990] and DMT/OFDM [Weinstein and Ebert, 1971, Peled and Ruiz, 1980, Cimini, Jr., 1985, Bingham, 1990, Cioffi, 1991, Ruiz et al., 1992, Zou and Wu, 1995]. Other proposed structures include Discrete Hartley Transform (DHT) basis vectors [Dudevoir, 1989] and the M-band wavelet transform in Discrete Wavelet MultiTone (DWMT) modulation [Tzannes et al., 1994].

2.   PARTITIONING FOR VECTOR CODING

The matrix H of size $N \times (N + \nu)$ in (2.2) has a Singular Value Decomposition (SVD)

$$H = U[\Lambda \vdots \mathbf{0}_{N,\nu}]V^* \qquad (2.7)$$

where U is an $N \times N$ unitary matrix, V is an $(N+\nu) \times (N+\nu)$ unitary matrix, and $\mathbf{0}_{N,\nu}$ is an $N \times \nu$ matrix of zeros. Λ is an $N \times N$ diagonal matrix with singular values λ_k, $k = 0, \ldots, N - 1$ along the diagonal. Vector Coding creates a set of N parallel independent channels by choosing as transmit basis vectors $\mathbf{m}_k$, the first N rows of V, i.e. $M = V$, and as receive basis vectors $\mathbf{f}_k$, the rows of U^*, i.e. $F = U$. With these substitutions in (2.5) and letting $\hat{\mathbf{X}}^m = [\mathbf{X}^{m*} \vdots \mathbf{0}_{1,\nu}]^*$,

$$
\begin{aligned}
\mathbf{Y}^m &= U^* H V \hat{\mathbf{X}}^m + U^* \mathbf{n}^m & (2.8) \\
&= U^* U[\Lambda \vdots \mathbf{0}_{N,\nu}]V^* V \hat{\mathbf{X}}^m + U^* \mathbf{n}^m & (2.9) \\
&= [\Lambda \vdots \mathbf{0}_{N,\nu}]\hat{\mathbf{X}}^m + U^* \mathbf{n}^m & (2.10) \\
&= \Lambda \mathbf{X}^m + \mathbf{N}^m & (2.11)
\end{aligned}
$$

Since U is unitary, the noise vector $\mathbf{N}^m$ is also additive white Gaussian with variance per dimension identical to $\mathbf{n}^m$. Equation (2.11) can also be expressed as N independent channels just as for the memoryless case in (2.6)

$$Y_k^m = \lambda_k X_k^m + N_k^m, \quad k = 0, \ldots, N - 1 \qquad (2.12)$$

This derivation shows that vector coding avoids the Inter-Symbol Interference (ISI) by appending zeros to $\mathbf{X}^m$ in (2.8) and by using the right and left singular vectors of the Toeplitz channel matrix H as transmit and receive vectors. Although it can be shown that Vector Coding has the maximum Signal to Noise Ratio (SNR) for any discrete channel partitioning [Cioffi, 2000b], it is not used in practical applications due to the complexity involved in computing the SVD.

3. PARTITIONING FOR DMT AND OFDM

Discrete MultiTone (DMT) and Orthogonal Frequency Division Multiplexing (OFDM) are the most common channel partitioning methods. By adding a restriction to the transmit sequence, the complexity of the transmitter and receiver is much lower than the Vector Coding case. Both DMT and OFDM use the same partitioning matrices, they only differ in the computation of the data vector $\mathbf{X}^m$. For OFDM all the components of the data vector $X_k^m, k = 0, \ldots, N-1$, i.e. all the subchannels are chosen from the same distribution. DMT modulators optimize the amount of energy $\mathcal{E}_k$ and also the amount of bits b_k in each subchannel to maximize the overall performance over a given channel. This optimization is called loading and will be described in more detail in Section 4. OFDM is used for broadcast applications since the data is demodulated by multiple receivers and the loading cannot be optimized to any particular user. OFDM is also used for point-to-point transmission in fast time-varying wireless environments where the receiver cannot feedback to the transmitter the optimal bits and energies. OFDM is the modulation format chosen for Digital Audio Broadcast (DAB) [ETSI, 1995] and Digital Video Broadcast (DVB) [EN300744, 1997, Reimers, 1998] and for second generation HIgh PErformance Radio Local Area Networks (HIPERLAN). DMT is used for point-to-point applications, usually on slowly time-varying channels, such as telephone lines [Chow et al., 1991a, Chow et al., 1991b]. DMT has been chosen as the modulation format for Asymmetric Digital Subscriber Line (ADSL), twisted pair telephone lines standard (ANSI T1.413 [ANSI, 1995] and ITU G.992 [ITU, 1999]) and is also a major contender in the ongoing Very high data rate Digital Subscriber Line (VDSL) standard.

To achieve channel partitioning, DMT/OFDM forces the modulated transmit vector $\mathbf{x}^m$ to satisfy the constraint

$$x_{-k}^m = x_{N-k}^m, \quad k = 1, \ldots, \nu \quad \forall m \tag{2.13}$$

i.e. the transmit vector is extended by copying the last ν samples of the multicarrier symbol at the beginning of the symbol. Replicating the last

ν samples at the beginning of the symbol is called Cyclic Prefix (CP) insertion. By adding the CP, (2.1) and (2.2) can be rewritten as,

$$\begin{bmatrix} y_{N-1} \\ \vdots \\ y_0 \end{bmatrix} = \begin{bmatrix} h_0 & h_1 & \ldots & h_\nu & 0 & 0 & \ldots & 0 \\ 0 & h_0 & h_1 & \ldots & h_\nu & 0 & \ldots & 0 \\ \vdots & \ddots & \ddots & \ddots & \ddots & \ddots & & \vdots \\ 0 & \ldots & 0 & h_0 & h_1 & \ldots & & h_\nu \\ h_\nu & 0 & \ldots & 0 & h_0 & \ldots & & h_{\nu-1} \\ \vdots & \ddots & \ddots & \ddots & \ddots & \ddots & & \vdots \\ h_1 & \ldots & h_\nu & 0 & \ldots & & 0 & h_0 \end{bmatrix} \begin{bmatrix} x_{N-1} \\ \vdots \\ x_0 \end{bmatrix} + \begin{bmatrix} n_{N-1} \\ \vdots \\ n_0 \end{bmatrix}$$

and more compactly as,

$$\mathbf{y} = \hat{H}\mathbf{x} + \mathbf{n} \tag{2.14}$$

where the matrix description of the channel $\hat{H}$, is a square $N \times N$ circulant matrix. For this special case, the SVD of $\hat{H}$ is much simpler to compute.

The Discrete Fourier Transform (DFT) of an N-dimensional vector $\mathbf{w} = [w_0 \ldots w_{N-1}]^T$ is also an N-dimensional vector $\mathbf{W} = [W_0 \ldots W_{N-1}]^T$ with components given by,

$$W_k = \frac{1}{\sqrt{N}} \sum_{n=0}^{N-1} w_n e^{-j\frac{2\pi}{N}kn}, \quad k = 0, \ldots, N-1 \tag{2.15}$$

Similarly, the Inverse Discrete Fourier Transform (IDFT) is given by

$$w_n = \frac{1}{\sqrt{N}} \sum_{k=0}^{N-1} W_k e^{j\frac{2\pi}{N}kn}, \quad n = 0, \ldots, N-1 \tag{2.16}$$

Equivalently, the DFT and IDFT can be written in matrix form as

$$\mathbf{W} = Q\mathbf{w} \tag{2.17}$$
$$\mathbf{w} = Q^*\mathbf{W} \tag{2.18}$$

where Q is the DFT orthonormal matrix with elements $q_{k,n} = \frac{1}{\sqrt{N}}e^{-j\frac{2\pi}{N}kn}$ and Q^* is the IDFT matrix. Using this notation, it can be shown [Cioffi, 2000b] that the circulant matrix $\hat{H}$ has eigen-decomposition

$$\hat{H} = Q^* \Lambda Q \tag{2.19}$$

where the diagonal values of Λ are $\lambda_k = H_k = DFT(\mathbf{h})$. Thus, choosing the columns of Q^* as transmit basis vectors, i.e. $M = Q^*$, and the rows of Q as receive basis vectors i.e. $F^* = Q$ the input-output relationship can be written as

$$Y_k^m = H_k X_k^m + N_k^m, \quad k = 0, \ldots, N-1 \qquad (2.20)$$

where again, the channel is partitioned into independent Additive White Gaussian Noise (AWGN) channels. The main advantage of the DMT and OFDM structure as compared to the standard Vector Coding, is that the DFT can be implemented with order $\mathcal{O}(N \log N)$ operations instead of $\mathcal{O}(N^2)$ operations for a general matrix multiplication. Therefore, transmitter ($\mathbf{x}^m = Q^* \mathbf{X}^m$) and receiver ($\mathbf{Y}^m = Q\mathbf{y}^m$) can be implemented very efficiently. The penalty to be paid for this large reduction in complexity is a slight performance degradation with respect to Vector Coding due to the cyclic prefix restriction, but this degradation is negligible when $N \gg \nu$.

4.   LOADING PRINCIPLES

Sections 2. and 3. showed that any finite length discrete-time channel can be partitioned into independent memoryless AWGN channels from (2.12) and (2.20). As described in Section 2. DMT modulators optimize the amount of energy $\mathcal{E}_k$ and also the amount of bits b_k in each subchannel to maximize the overall performance over a given channel. This operation is called loading and is introduced in this section along with several references.

As described by Shannon in two early papers that constitute the foundations of information theory [Shannon, 1948a, Shannon, 1948b], when the channel is memoryless, the channel can be characterized by a single value C called channel capacity, which quantifies the maximum error-free data rate achievable with unconstrained complexity and unlimited decoding delay. Mathematically,

$$C = \log_2(1 + SNR) \qquad (2.21)$$

where the Signal to Noise Ratio (SNR) is defined as

$$SNR = \frac{|H|^2 \mathcal{E}_X}{2\sigma^2} \qquad (2.22)$$

Here, H is the channel gain, $\mathcal{E}_X = E\{|X|^2\}$ denotes the two-dimensional (2D) symbol transmit power, $2\sigma^2 = E\{|n|^2\}$ is the receiver additive white Gaussian noise variance, and $E\{\cdot\}$ denotes expectation. From (2.20) and (2.21), the capacity of the k-th subchannel of a multicarrier system is given by

$$C_k = \log_2(1 + SNR_k) = \log_2\left(1 + \frac{|H_k|^2\mathcal{E}_{X,k}}{2\sigma_k^2}\right) \qquad (2.23)$$

$$= \log_2\left(1 + g_k\mathcal{E}_{X,k}\right) \qquad (2.24)$$

where $\mathcal{E}_{X,k} = E\{|X_k|^2\}$ denotes the 2D symbol power allocated to the k-th subchannel and $2\sigma_k^2 = E\{|N_k|^2\}$ is the additive Gaussian noise variance. For the case where the noise is white, $\sigma_k = \sigma_0$, $k = 1 \ldots N - 1$ we have N independent Gaussian channels in parallel. Assuming there is a common power constraint, $\sum_{k=0}^{N-1} \mathcal{E}_{X,k} \leq \mathcal{E}_{total}$, the optimal power distribution among the various subchannels can be computed so as to maximize the total capacity [Gallager, 1968, Cover and Thomas, 1991]. In practice, the transmission system must operate under finite complexity and delay constraints and can tolerate nonzero error rates. Moreover, in most multicarrier applications, data is transmitted by first mapping bits into complex symbols selected from a Quadrature Amplitude Modulated (QAM) constellation. Under these constraints, a useful approximation for computing the achievable data rate in an AWGN channel is given by the SNR gap approximation [Cioffi et al., 1995, Cioffi, 2000b, Cioffi, 2000a, Chow, 1993].

$$R = \log_2\left(1 + \frac{SNR}{\Gamma(\mathcal{C}, P_e)}\right) \qquad (2.25)$$

where the SNR gap term $\Gamma(\mathcal{C}, P_e)$ is a function of the coding scheme $\mathcal{C}$ and the desired error rate P_e. Therefore, the number of bits assigned to the k-th multicarrier subchannel is:

$$b_k = \log_2\left(1 + \frac{SNR_k}{\Gamma(\mathcal{C}, P_e)}\right) \qquad (2.26)$$

and the aggregate rate over all subchannels is given by:

$$R = \frac{B}{T} = \frac{1}{T}\sum_{k=0}^{N-1} b_k = \frac{1}{T}\sum_{k=0}^{N-1} \log_2\left(1 + \frac{SNR_k}{\Gamma(\mathcal{C}, P_e)}\right) \qquad (2.27)$$

where T is the multicarrier symbol duration. The optimal energy allocation under a common power constraint, or fixed data rate constraint is called *waterfilling*. Unfortunately, ideal waterfilling leads to real values for the bit assignment vector $\mathbf{b} = [b_0 \ldots b_{N-1}]^T$. Since b_k can only take on a discrete set of values, typically positive integers and occasionally simple fractions, discrete waterfilling algorithms are needed [Hughes-Hartogs, 1989, Chow, 1993, Campello de Sousa, 1998, Campello de Sousa, 1999]. In many applications, the bits transmitted may have different target BER (e.g. GSM voice coders). In other occasions, we may desire to transmit two data streams with different reliability requirements simultaneously (e.g. data and voice) over the same channel. In these cases, the value of the SNR gap $\Gamma(\mathcal{C}_k, P_{e,k})$ can be different for each bit stream. Multi-gap/multi-service and multi-user loading algorithms are described in [Hoo et al., 1998b, Hoo et al., 1998a, Hoo et al., 1999] for both the real valued and discrete bit assignments.

Chapter 3

PEAK TO AVERAGE RATIO

C HAPTER 2 BRIEFLY DESCRIBED some of the basics of multicarrier transmission that have justified its increased presence in high-performance communications systems. Unfortunately, one particular problem with multicarrier signals, which is often cited as the major drawback of DMT/OFDM transmission, is its large envelope fluctuations. Since all practical transmission systems are peak-power limited, designing our system to operate in a perfectly linear region often implies operating at average power levels way below the maximum power available. In practice, to avoid operating the amplifiers with extremely large back-offs, occasional saturation of the power amplifiers or clipping in the Digital-to-Analog Converters (DAC) must be allowed. This additional nonlinear distortion creates inter-modulation distortion that increases the bit error rate in standard linear receivers, and also causes spectral widening of the transmit signal that increases adjacent-channel interference to other users.

The main focus of this chapter is to introduce the problem of Peak to Average Power Ratio (PAR) and provide most of the necessary background for understanding the subsequent chapters.

1. MULTICARRIER SIGNALS

A DMT/OFDM transmit signal is the sum of N, independent, Quadrature Amplitude Modulated (QAM) sub-signals or tones, each with equal bandwidth and frequency separation $1/T$, where T is the time duration of the multicarrier symbol [Bingham, 1990]. The continuous-time baseband representation of a single multicarrier symbol is given by

$$x^m(t) = \frac{1}{\sqrt{N}} \sum_{k=-\frac{N}{2}}^{\frac{N}{2}-1} X_k^m e^{j2\pi kt/T} w(t), \qquad (3.1)$$

where m is the symbol index, $w(t)$ is a rectangular window (nominally of height 1 over the interval $[0, T]$) and X_k^m is the QAM value of the *k-th* subsymbol or tone. To simplify the equalizer design in the presence of finite-length channel multipath, i.e. Finite Impulse Response (FIR) ISI, multicarrier systems insert a Cyclic Prefix[1] (CP) before every multicarrier symbol. As described in Chapter 2 in the context of Channel Partitioning, the equalizer then becomes a simple scaling as long as the duration of the combined effect of channel multipath plus transmit and receiver filtering is shorter than the length of the CP, T_{CP}. The cyclic prefix is simply a periodic extension of the symbol over the interval $[-T_{CP}, 0]$ resulting in a symbol of length $[-T_{CP}, T]$. Equation (3.1) thus becomes,

$$x_{CP}^m(t) = \frac{1}{\sqrt{N}} \sum_{k=-\frac{N}{2}}^{\frac{N}{2}-1} X_k^m e^{j2\pi kt/T} w_{CP}(t), \qquad (3.2)$$

where $w_{CP}(t)$ is a rectangular window of height 1 over the interval $[-T_{CP}, T]$. For continuous data transmission, the transmitter sends these symbols sequentially:

$$x_C(t) = \sum_{m=-\infty}^{\infty} x_{CP}^m(t - m(T_{CP} + T)). \qquad (3.3)$$

This transmit signal is not bandlimited due to the $sinc(f(T + T_{CP}))$ behavior exhibited by the rectangular windowing function $w_{CP}(t)$, and is typically followed by a filter. Moreover, with this representation, computing $x_C(t)$ requires a Continuous Time Fourier Transform (CTFT), which is very hard to implement with analog components and can only be approximated with digital hardware. Therefore in practice, complex baseband DMT/OFDM signals are typically generated by using an Inverse Discrete Fourier Transform (IDFT) as described by the block diagram in Figure 3.1.

---

[1]Some authors use the term *guard interval* instead.

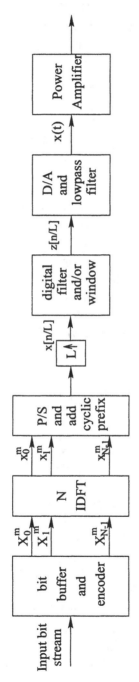

Figure 3.1: DMT/OFDM Transmitter Block Diagram.

The *m-th* block of encoded bits is mapped into the complex-valued DMT/OFDM vector of QAM or MPSK constellation points $\mathbf{X}^m = [X_0^m \ldots X_{N-1}^m]^T$, which is then transformed via an IDFT into the T/N-spaced discrete-time vector $\mathbf{x}^m = x^m[n] = [x_0^m \ldots x_{N-1}^m]^T = IDFT(\mathbf{X}^m)$, i.e.

$$\mathbf{x}^m = x^m[n] \quad = \quad \frac{1}{\sqrt{N}} \sum_{k=0}^{N-1} X_k^m e^{j2\pi kn/N} w[n] \qquad (3.4)$$

$$= \quad \frac{1}{\sqrt{N}} \sum_{k=-\frac{N}{2}}^{\frac{N}{2}-1} X_k^m e^{j2\pi kn/N} w[n], \qquad (3.5)$$

where $w[n]$ is the discrete-time rectangular window of height 1 over the interval $[0, N-1]$. Moreover, in (3.5) the index k of the QAM subsymbol X_k^m is computed modulo N.

In this book, the discrete-time indexing $[n]$ denotes Nyquist rate samples. Since oversampling may be needed in practical designs, we will introduce the notation $x[n/L]$ to denote oversampling by a factor L. Several different oversampling strategies of $x^m[n]$ can be defined. From now on, the oversampled IDFT output will refer to oversampling of (3.5), which is expressed as follows:

$$x^m[n/L] = \frac{1}{\sqrt{N}} \sum_{k=-\frac{N}{2}}^{\frac{N}{2}-1} X_k^m e^{j2\pi kn/NL} w[n/L] \qquad (3.6)$$

$$= \quad \frac{w[n/L]}{\sqrt{N}} \left(\sum_{k=0}^{\frac{N}{2}-1} X_k^m e^{j2\pi kn/NL} + \sum_{k=NL-\frac{N}{2}}^{NL-1} X_{k-N(L-1)}^m e^{j2\pi kn/NL} \right) (3.7)$$

$$= \quad IDFT(\sqrt{L}[X_0^m \ldots X_{\frac{N}{2}-1}^m \underbrace{0 \ldots 0}_{N(L-1)} X_{\frac{N}{2}}^m \ldots X_{N-1}^m]^T) \qquad (3.8)$$

$$= \quad IDFT(\sqrt{L}\,\mathbf{X}_L^m) \qquad (3.9)$$

where $w[n/L]$ is the discrete rectangular window of height 1 over the interval $n \in [0, NL-1]$ and $\mathbf{X}_L^m$ is the L times oversampled equivalent QAM vector, generated by *zero padding* $\mathbf{X}^m$ with $N(L-1)$ zeros.

With these definitions, the CTFT (3.1) and DFT (3.6) time-domain representations of $\mathbf{X}^m$ are closely related as it's easy to show that $x^m(nT/NL) = x^m[n/L]$. As mentioned earlier, to simplify ISI mitigation at the receiver, the transmitter must add a CP to each vector $\mathbf{x}^m$

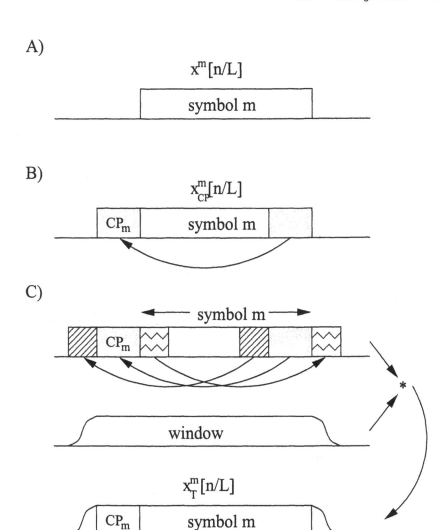

Figure 3.2: Different Multicarrier Symbols: A) Basic, B) With CP and C) With CP and windowed extended CP and CS.

(see Figure 3.2,B)

$$x_{CP}^m[n/L] = \frac{1}{\sqrt{N}} \sum_{k=-\frac{N}{2}}^{\frac{N}{2}-1} X_k^m e^{j2\pi kn/NL} w_{CP}[n/L], \qquad (3.10)$$

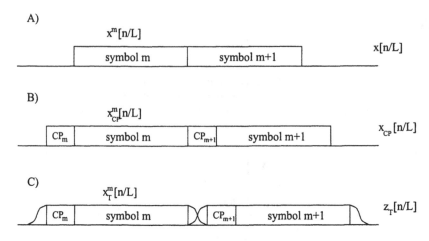

Figure 3.3: Multicarrier Signals from Figure 3.2.

where $w_{CP}[n/L]$ is the rectangular window over the interval $[-\nu L, NL - 1]$. The length of the discrete CP is νL and its equivalent continuous-time domain span is $T_{CP} = \nu T/N$. These extended vectors are then transmitted sequentially as described by (see Figure 3.3,B):

$$x_{CP}[n/L] = \sum_{m=-\infty}^{\infty} x_{CP}^m[(n - m(N + \nu)L)/L] \qquad (3.11)$$

This sequence is not bandlimited because of the slow decay with increasing frequency of the Fourier transform of a rectangular window. The solid curve ($\beta = 0$) in Figure 3.4 plots the Power Spectral Density (PSD) of $x[n/4]$ for a real baseband case with $N = 512$ and $\nu = 32$. From this figure, the PSD decays very slowly and is only slightly below -40 dB at a frequency four times higher than the nominal bandwidth. If the transmitter must meet strict adjacent channel interference requirements, some degree of filtering or smooth windowing will be needed. This filtering can be performed on the discrete-time samples prior to the DAC and/or on the continuous-time output of the DAC (see (3.15)). If the spectral confinement is performed on the discrete time sequence $x[n/L]$ with a digital filter $p[n/L]$, the filtered signal is expressed as:

$$z_F[n/L] = x[n/L] * p[n/L], \qquad (3.12)$$

where $*$ denotes convolution.

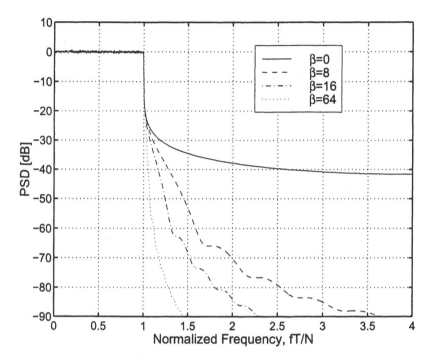

Figure 3.4: PSD of the discrete-time multicarrier signal
in (3.14) for $N = 512$, $L = 4$ and raised
cosine windows with different roll-off lengths.

Alternatively, the spectral properties can be improved by smoother windowing instead of filtering. In general, windowing the multicarrier symbols destroys the orthogonality between the tones, which leads to Inter-Carrier Interference (ICI). To avoid loss of orthogonality, the transmitter design can increase the cyclic extension and modify the rectangular windowing sequence as described in [Weinstein and Ebert, 1971, Pauli and Kuchenbecker, 1998] (see Figure 3.2,C). For this case, the transmit sequence becomes

$$x_T^m[n/L] = \frac{1}{\sqrt{N}} \sum_{k=-\frac{N}{2}}^{\frac{N}{2}-1} X_k^m e^{j2\pi kn/NL} w_T[n/L], \qquad (3.13)$$

where the tapered window sequence $w_T[n/L]$, $n = -(\nu+\beta/2)L, \dots, (N+\beta/2)L - 1$ is constant over the interval $[-\nu L, NL - 1]$ and tapers off on the remaining βL samples, with $\beta L/2$ tapered samples each at the beginning and end of the symbol (see Figure 3.2,C). These smoothed transmit-sample vectors are then transmitted sequentially as described by (see Figure 3.3,C):

$$z_T[n/L] = \sum_{m=-\infty}^{\infty} x_T^m[(n - m(N + \nu + \beta/2)L)/L] \qquad (3.14)$$

The effect of windowing on the discrete-time multicarrier symbols is shown in Figure 3.4 for $N = 512$, $L = 4$. The window used is a raised cosine which is constant over the middle $(N + \nu)L$ samples and with roll-off of length $\beta L/2$ on each side. As seen in the figure the PSD can be reduced very significantly for small values of β. The maximum amount of overhead, or penalty from windowing relative to the symbol length is β/N. In fact, the last $\beta L/2$ tapered samples of a symbol can be overlapped with the first $\beta L/2$ tapered samples of the next symbol as shown in Figure 3.3,C. In this case, the overhead is only $\beta/2N$. Thus, for the example in Figure 3.4, the overhead for the $\beta = 16$ case is 1.5%.

The sequences in (3.12) or (3.14) are then passed through a DAC and an analog filter to produce the continuous-time signal:

$$x_D(t) = \sum_{k=-\infty}^{\infty} z[k/L]p(t - \frac{kT}{NL}) \qquad (3.15)$$

where N/T is Nyquist rate and $p(t)$ includes the effects of the DAC response and any subsequent analog filtering prior to amplification.

2.   PEAK TO AVERAGE RATIO

Since the cost of transceiver components depends on the dynamic range of the signals in (3.1)-(3.15), it is of interest to characterize these signals.

It is has been shown in the literature that when N is large, the transmit multicarrier signals in (3.1)-(3.15) can be approximately modeled by truncated Gaussian random processes with zero mean. This section will review some of these concepts.

Typically, the PAR is used to quantify the envelope excursions of the signal[2]. The PAR of a signal x_τ, where τ is used to represent both the

---

[2]Other authors use the closely related parameter called Crest Factor (CF) where $CF = \sqrt{PAR}$

continuous-time index n and discrete-time index t, is defined as:

$$PAR\{x_\mathcal{T}, \mathcal{T}\} = \frac{\max\limits_{\tau \in \mathcal{T}} |x_\tau|^2}{E\{|x_\tau|^2\}} \qquad (3.16)$$

Here, $\max\limits_{\tau \in \mathcal{T}} |x_\tau|^2$ denotes the maximum instantaneous power, $E\{|x_\tau|^2\}$ denotes the average power of the signal and $\mathcal{T}$ is the interval over which the PAR is evaluated. Moreover, x_τ can be replaced by the discrete-time signals $x^m[n/L]$, $x_{CP}^m[n/L]$, $x_T^m[n/L]$, $x_{CP}[n/L]$, $z_F[n/L]$, $z_T[n/L]$, or the continuous-time waveforms $x^m(t)$, $x_{CP}^m(t)$ and $x_D(t)$ as defined in the previous section. Throughout this book, the PAR will refer to the baseband PAR. For passband transmission, the transmit multicarrier signal is modulated onto a carrier frequency f_c,

$$
\begin{aligned}
x_{PB}(t) &= \Re\{x(t)e^{j2\pi f_c t}\} & (3.17)\\
&= \Re\{x(t)\}\cos(j2\pi f_c t) - j\Im\{x(t)\}\sin(j2\pi f_c t) & (3.18)\\
&= x_I(t)\cos(j2\pi f_c t) - jx_Q(t)\sin(j2\pi f_c t) & (3.19)
\end{aligned}
$$

Since the carrier frequency is usually much higher than the signal bandwidth, i.e. $f_c \gg N/T$, from (3.17), the maximum of the modulated signal is the same as the maximum of the baseband signal, i.e.

$$\max|x_{PB}(t)| \approx \max|x(t)| \qquad (3.20)$$

For QAM modulation $E\{|x_I(t)|^2\} = E\{|x_Q(t)|^2\} = \frac{1}{2}E\{|x(t)|^2\}$. From (3.19) the average powers can be written

$$E\{|x_{PB}(t)|^2\} = \frac{1}{2}E\{|x_I(t)|^2\} + \frac{1}{2}E\{|x_Q(t)|^2\} = \frac{E\{|x(t)|^2\}}{2} \qquad (3.21)$$

From (3.20) and (3.21) the passband PAR is roughly twice (3 dB higher) than the baseband PAR, i.e.

$$
\begin{aligned}
PAR\{x_{PB}(t)\} &= \frac{\max\limits_{t \in \mathcal{T}} |x_{PB}(t)|^2}{E\{|x_{PB}(t)|^2\}} & (3.22)\\
&\approx \frac{\max\limits_{t \in \mathcal{T}} |x(t)|^2}{E\{|x(t)|^2\}/2} = 2PAR\{x(t)\} & (3.23)
\end{aligned}
$$

The PAR in (3.16) can be computed exactly for some of the DMT/OFDM signals defined in Section 1. For example, the PAR of $x^m[n/L]$ defined in (3.6) can be easily computed. The peak power is

$$\max_n |x^m[n/L]|^2 \;=\; \frac{1}{N}\left|\sum_{k=-\frac{N}{2}}^{\frac{N}{2}-1} X_k^m e^{j2\pi kn/NL}\right|^2 \tag{3.24}$$

$$\leq \frac{1}{N}\left(\sum_{k=0}^{N-1} \max |X_k^m|\right)^2 \tag{3.25}$$

Using Parseval's relationship the Average Power is

$$E\{|x^m[n/L]|^2\} = \frac{1}{N}\sum_{k=0}^{N-1} E\{|X_k^m|^2\} \tag{3.26}$$

from which the PAR can be computed. For the case where all tones share the same constellation, which is the case for all OFDM systems, the PAR has a simple maximum:

$$PAR\{x^m[n/L]\} \leq N\frac{\max |X_k|^2}{E\{|X_k|^2\}} \tag{3.27}$$

where equality is achieved for example at $n = 0$ when all QAM subsymbols have the same phase $\arg\{X_0^m\} = \arg\{X_k^m\}$, $k = 1,\dots,N-1$, and one of the QAM points with largest amplitude is selected. Moreover, since the CP is just a replica of the last νL samples of $x^m[n/L]$, it does not change the average or peak power and therefore:

$$PAR\{x_{CP}^m[n/L]\} = PAR\{x^m[n/L]\} \leq N\frac{\max |X_k|^2}{E\{|X_k|^2\}} \tag{3.28}$$

Since $x_{CP}[n/L]$ in (3.11) is constructed from non-overlapping sequential transmissions of the symbols $x_{CP}^m[n/L]$, we have:

$$PAR\{x_{CP}[n/L]\} = \max_m PAR\{x_{CP}^m[n/L]\} = N\frac{\max |X_k|^2}{E\{|X_k|^2\}} \tag{3.29}$$

which shows that the PAR grows linearly with the number of tones and is proportional to the PAR of the constellation. This PAR expression also holds for $z_T[n/L]$ when the real valued tapered window $w_T[n/L]$ satisfies the symmetry property

$$w_T[n/L] + w_T[(n + (N + \nu + \beta/2)L)/L] = 1, \tag{3.30}$$

over the range $n = -(\nu + \beta/2)L, \ldots, -\nu L$ and

$$0 \leq w_T[n/L] \leq 1, \forall n. \tag{3.31}$$

This can be easily understood from observing Figure 3.3,C. Obviously, over the non-overlapping sections of $z_T[n/L]$ there is no change in the peak or average power. For notational simplicity, let's consider the overlapping sections in the interval $n = -(\nu + \beta/2)L, \ldots, -\nu L$, i.e. symbols $m = -1$ and $m = 0$

$$
\begin{aligned}
|z_T[n/L]| &= \left| x_T^{-1}[(n + (N + \nu + \beta/2)L)/L] + x_T^0[n/L] \right| \tag{3.32} \\
&\leq \max_m |x^m[n/L]| w_T[n/L] \\
&\quad + \max_m |x^m[n/L]| w_T[(n + (N + \nu + \beta/2)L)/L] \tag{3.33} \\
&= \max_{n,m} |x^m[n/L]| \tag{3.34}
\end{aligned}
$$

Similarly, assuming symbols $m = -1$ and $m = 0$ are uncorrelated

$$
\begin{aligned}
E\{|z_T[n/L]|^2\} &= \\
&= E\left\{ \left| x_T^{-1}[(n + (N + \nu + \beta/2)L)/L] + x_T^0[n/L] \right|^2 \right\} \tag{3.35} \\
&= E\{|x^m[n/L]|^2\} |w_T[n/L]|^2 \\
&\quad + E\{|x^m[n/L]|^2\} |w_T[(n + (N + \nu + \beta/2)L)/L]|^2 \tag{3.36} \\
&= E\{|x^m[n/L]|^2\} \Big(|w_T[n/L]|^2 \\
&\quad + |w_T[(n + (N + \nu + \beta/2)L)/L]|^2 \Big) \tag{3.37}
\end{aligned}
$$

where the expected value is taken over the data symbol vector $\mathbf{X}$. Since $\frac{1}{2} \leq |w_T[n/L]|^2 + |w_T[(n + (N + \nu + \beta/2)L)/L]|^2 \leq 1$, we have

$$\frac{1}{2} E\{|x^m[n/L]|^2\} \leq E\{|z_T[n/L]|^2\} \leq E\{|x^m[n/L]|^2\} \tag{3.38}$$

Since the right inequality is strict only when $|w_T[n/L]| < 1$ (tapering section), the average power is lower and upper bounded as follows

$$\frac{P - \beta L/4}{P} \frac{1}{P} \sum_{n=0}^{P-1} E\{|x^m[n/L]|^2\} \leq \tag{3.39}$$

$$\leq \frac{1}{P} \sum_{n=0}^{P-1} E\{|z_T[n/L]|^2\} \leq \frac{1}{P} \sum_{n=0}^{P-1} E\{|x^m[n/L]|^2\} \tag{3.40}$$

where $P = (N + \nu + \beta/2)L$ is the CP-CS extended multicarrier symbol length. Typically $\beta << N$ and

$$E\{|z_T[n/L]|^2\} \approx E\{|x^m[n/L]|^2\} \qquad (3.41)$$

Since linear transmission of these multicarrier signals requires linear operation over the range $(-\max_n |x_n|, \max_n |x_n|)$, the average transmit power for zero distortion is at most $1/N$ times the maximum available power. As practical values of N are on the order of $N = 256$ [ANSI, 1995] to $N = 8196$ [EN300744, 1997], the consequent power efficiency is below 1%. Therefore, to increase the efficiency, some type of PAR reduction method is desirable. Although

$$\max_m PAR\{x^m[n/L]\} \geq N \qquad (3.42)$$

in practice, $PAR\{x^m[n/L]\} << N$ for most symbols. The following section studies the statistical distribution of the multicarrier symbol PAR.

3.   STATISTICAL PROPERTIES OF MULTICARRIER SIGNALS

This section describes only critically sampled (i.e. $L = 1$) discrete-time multicarrier signals while Section 4. is devoted to oversampled and continuous-time multicarrier signals. Let's first consider the simple case where all the subsymbols X_k^m are i.i.d. complex Gaussian random variables with zero mean and unit variance. From (3.4), it can be shown that the symbol samples $x^m[n]$ are also i.i.d. Gaussian random variables with zero mean and unit variance. As each sample $x^m[n]$ is a linear combination of Gaussian random variables, it is also a Gaussian random variable. Since the IDFT operation as defined in (3.4) is an orthogonal transformation, each sample $x^m[n]$ has unit variance. Moreover, the independence also follows from the row orthogonality. Since

$$E\{x^m[r](x^m[s])^*\} = \frac{1}{N}E\left\{\left(\sum_{l=0}^{N-1} X_l^m e^{j2\pi lr/N}\right)\left(\sum_{k=0}^{N-1} X_k^m e^{j2\pi ks/N}\right)^*\right\}$$

$$(3.43)$$

and uncorrelated Gaussian variables are also independent, we have:

$$E\{x^m[r](x^m[s])^*\} = \frac{1}{N}\sum_{k=0}^{N-1}E\{X_k^m(X_k^m)^*\}e^{j2\pi kr/N}e^{-j2\pi ks/N} \quad (3.44)$$

$$= \delta[r-s] \quad (3.45)$$

In most practical applications, the data is randomized prior to modulation and the subsymbols X_k^m can be approximated as independent discrete uniform random variables, typically MQAM, MPSK or APSK. For these cases, the symbol samples $x^m[n]$ are linear combination of N discrete uniform random variables. For the OFDM case, all subsymbols X_k^m are chosen from the same constellation and thus the N discrete uniform random variables are i.i.d. Since the subsymbols X_k^m are independent, from (3.43-3.45) the symbol samples $x^m[n]$ are still uncorrelated. Moreover, N is typically large. Thus, from the Central Limit Theorem, the symbol samples $x^m[n]$ are approximately Gaussian. This leads to the common assumption that for large values of N, the symbol samples are approximately i.i.d. random variables [Müller et al., 1997, Wulich et al., 1998]. With this assumption, the Cumulative Distribution Function (CDF) of the random variable $PAR\{x^m[n]\}$ has a simple closed form distribution:

$$Prob\left\{PAR\{x^m[n]\} < \gamma^2\right\} = Prob\left\{\frac{|x^m[0]|^2}{E\{|x^m[n]|^2\}} < \gamma^2, \cdots\right.$$

$$\left.\cdots, \frac{|x^m[N-1]|^2}{E\{|x^m[n]|^2\}} < \gamma^2\right\} \quad (3.46)$$

$$= \left(Prob\left\{\frac{|x^m[n]|^2}{E\{|x^m[n]|^2\}} < \gamma^2\right\}\right)^N \quad (3.47)$$

For real (baseband) multicarrier symbols the CDF of $PAR\{x^m[n]\}$ is

$$Prob\{PAR\{x^m[n]\} < \gamma^2\} = \left(1 - 2Q(\gamma)\right)^N \quad (3.48)$$

and thus the Complementary Cumulative Distribution Function (CCDF) is

$$Prob\{PAR\{x^m[n]\} > \gamma^2\} = 1 - \left(1 - 2Q(\gamma)\right)^N \quad (3.49)$$

For the complex multicarrier case the symbol CDF of $PAR\{x^m[n]\}$ is

$$Prob\{PAR\{x^m[n]\} < \gamma^2\} = \left(1 - exp(-\gamma^2)\right)^N \quad (3.50)$$

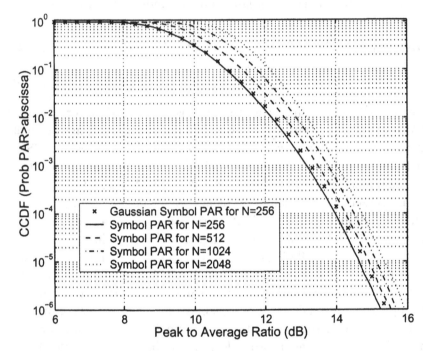

Figure 3.5: CCDF of $PAR\{x^m[n]\}$ for
$N = 256, 512, 1024, 2048$.

Figure 3.5 shows the CCDF of $PAR\{x^m[n]\}$ for different values of N, in particular $N = 256$, 512, 1024 and 2048 generated by Monte Carlo simulation. For comparison, we include the PAR for an i.i.d. Gaussian vector of size $N = 256$. This plot shows that the Gaussian assumption is good for predicting the distribution of $PAR\{x^m[n]\}$. However, as shown in [Tellado and Cioffi, 2000b], this approximation is not accurate when predicting mutual information loss from PAR reduction algorithms. Moreover, from the figure, it can be inferred that 99.9999% of the symbols have $PAR\{x^m[n/L]\} \leq 40$ (16 dB) $<< N$ and the statistical PAR does not increase significantly with increased symbol size, N. Taking the ADSL case as an example ($N = 512$, $T = 250$ μsec), although the maximum PAR is $10\log(512) = 27$ dB, only once every 4 minutes ($10^6 \times 250$ μsec) does the PAR exceed $10\log(35) = 15.5$ dB. Moreover, from the Gaussian distribution, a 2 dB increase in allowable PAR will increase the rate from once every 4 minutes to once every few days.

Thus, the often quoted PAR value of N for multicarrier signals is extremely pessimistic for appropriately randomized data. Although the PAR for practical systems is typically below 15-17 dB most of the time, this value is still high and PAR reduction methods are still of great interest. There are many different alternatives in the literature, many of which are summarized in Section 8. The subsequent chapters are devoted to describing our novel methods for reducing the PAR in multicarrier transmission.

4. BOUNDS ON CONTINUOUS-TIME PAR USING DISCRETE-TIME SAMPLES

The PAR of the discrete-time sequences typically determines the complexity of the digital circuitry in terms of the number of bits necessary to achieve a desired Signal to Quantization Noise for both the digital signal processing and the DAC. In practice, we are often more concerned with reducing the PAR of the continuous-time signal, since the cost and power dissipation of the analog components often dominates. As most PAR reduction methods can only be implemented on the discrete-time signals, this section describes the relationship between discrete-time and continuous-time PAR. The following discussion assumes that the over-sampled IDFT outputs are generated as described by (3.4)-(3.11). For this case, $x^m[n/L] = x^m(nT/NL) = x^m(t)_{t=nT/NL}$ and thus:

$$\max_n |x^m[n/L]| = \max_n |x^m(t)|_{t=nT/NL} \leq \max_t |x^m(t)| \qquad (3.51)$$

Since $E\{|x^m[n/L]|^2\} = E\{|x^m(t)|^2\}$, the PAR satisfies

$$PAR\{x^m[n/L]\} \leq PAR\{x^m(t)\} \qquad (3.52)$$

i.e. the continuous-time PAR is larger than or equal to the discrete-time PAR. Let's consider a set of oversampling rates $\mathcal{L} = \{L_1, \ldots, L_R\}$ such that

$$L_{r+1} = K_r L_r, \quad r = 1, \ldots, R-1 \qquad (3.53)$$

where the set of indexes $\mathcal{K} = \{K_1, \ldots, K_R\}$ are strictly positive integers. For this case the maxima can be written

$$\max_n |x^m[n/L_r]| \leq \max_n |x^m[n/L_{r+1}]|, \quad r = 1, \ldots, R-1 \qquad (3.54)$$

since the sequence of points $x^m[n/L_r]$ is included in the sequence of points $x^m[n/L_{r+1}]$. Since the average power does not change, the PAR must be

$$PAR\{x^m[n/L_r]\} \leq PAR\{x^m[n/L_{r+1}]\} \tag{3.55}$$

and therefore, increasing the oversampling rate by integer amounts increases the symbol PAR. In general, if the oversampling rates $\mathcal{L} = \{L_1,\ldots,L_R\}$ satisfy $L_1 \leq \cdots \leq L_r \leq \cdots \leq L_R$, the samples will be different and the ordering will not hold on a symbol by symbol basis, but the CCDF satisfy

$$Prob\{PAR\{x^m[n/L_r]\} \geq \gamma\} \leq Prob\{PAR\{x^m[n/L_{r+1}] \geq \gamma\} \tag{3.56}$$

Since $x^m(t)$ is a continuous function over $(0,T)$, we have:

$$\lim_{L\to\infty} PAR\{x^m[n/L]\} = PAR\{x^m(t)\} \tag{3.57}$$

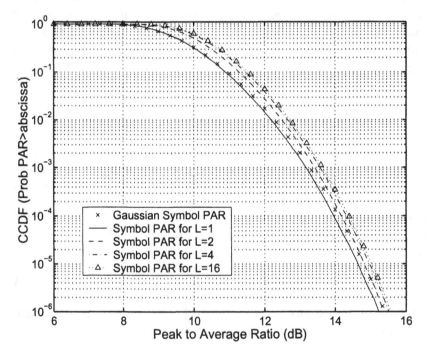

Figure 3.6: CCDF of $PAR\{x^m[n/L]\}$ for $L = 1, 2, 4, 16$.

Figure 3.6 is a plot of the CCDF of $PAR\{x^m[n/L]\}$ for $N = 256$ and $\mathcal{L} = \{1, 2, 4, 16\}$. As shown, the largest PAR CCDF increase happens when L is increased from $L = 1$ to $L = 2$ but does not increase significantly after $L = 4$. From the Central Limit Theorem, all the samples in $x^m[n/L]$ are approximately Gaussian distributed, since all the samples are generated from a large linear combination of well behaved random variables. As L increases, there are more samples to increase the PAR, but these extra samples are more correlated to their neighbors, resulting in smaller increases in PAR CCDF. Figure 3.6 might lead us to the conclusion that critical sampling (i.e. $L = 1$) can predict the continuous-time ($L \to \infty$) PAR within 1 dB. Unfortunately, this analysis does not hold in general when PAR reduction is performed on the multicarrier symbol. Figure 3.7 shows the PAR CCDF for an ideal PAR reduction method that achieves $PAR\{x^m[n]\} \le 10$ dB. This method simply selects those symbols with PAR below the 10 dB threshold and disregards the rest. For this case, the PAR has been reduced by 5.5 dB at 10^{-6} CCDF point for the critically sampled symbol. Unfortunately, the oversampled signal ($L = 2$) has only a 1 dB PAR reduction at 10^{-6} CCDF point. Therefore this ideal PAR reduction method operating on critically sampled symbols does a poor job at reducing the PAR of the oversampled and continuous-time signals. The rest of this section is devoted to studying the maximum difference between the discrete-time symbols and their oversampled and continuous-time equivalents.

Most of the results in this book can be extended to passband applications of DMT/OFDM (e.g. DAB or DVB), where the baseband sequences, $x[n/L]$, $z[n/L]$ and $x(t)$ are complex valued. However, to simplify the discussion, in some mathematical derivations, baseband transmission of DMT/OFDM is assumed, which means that x_τ must be real. Assuming N is even, for x_τ^m to be a real sequence, X_0^m and $X_{N/2}^m$ must be real, and X_k^m must satisfy $X_k^m = X_{-k}^m{}^*{}_{\bmod N} = X_{N-k}^m{}^*$ where * denotes complex conjugate. Thus, for this case, there are only $N/2$ independent complex-values in the QAM vector that must be specified. Similar constraints are necessary for N odd.

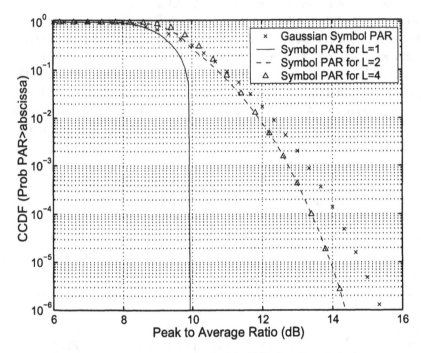

Figure 3.7: CCDF of $PAR\{x^m[n/L]\}$ for $L = 2, 4$ given $PAR\{x^m[n]\} \leq 10$ dB.

For the real-valued case the multicarrier symbol in (3.1) simplifies to[3]

$$x^m(t) = \frac{2}{\sqrt{N}} \sum_{k=1}^{\frac{N}{2}-1} [R_k^m \cos(2\pi kt/T) - I_k^m \sin(2\pi kt/T)] \qquad (3.58)$$

where $t \in (0, T)$. To simplify the following derivation, the symbol index m will be dropped. This signal is continuous and differentiable over $(0, T)$. The derivative of $x(t)$ is

$$x'(t) = \frac{-2}{\sqrt{N}} \sum_{k=1}^{\frac{N}{2}-1} \left[R_k \frac{2\pi k}{T} \sin(2\pi kt/T) + I_k \frac{2\pi k}{T} \cos(2\pi kt/T) \right] \qquad (3.59)$$

---

[3]The DC term (X_0^m) and the Nyquist term ($X_{-N/2}^m$) are set to zero for simplicity.

The maximum slope can be upper bounded as follows:

$$|x'(t)| \leq \frac{2}{\sqrt{N}} \sum_{k=1}^{\frac{N}{2}-1} \left[\frac{2\pi k}{T} \max(|R_k| + |I_k|) \right] \tag{3.60}$$

Assuming equal constellations in every tone with normalized power $E\{|X_k|^2\} = 1$ it can be shown that

$$\max |R_k| = \max |I_k| = \frac{(\sqrt{M}-1)d}{2} = \frac{\sqrt{M}-1}{2} \sqrt{\frac{6}{M-1}} \leq \sqrt{\frac{3}{2}} \tag{3.61}$$

Using this bound, the slope can be upper bounded by

$$|x'(t)| \leq \frac{4\pi\sqrt{6}}{T\sqrt{N}} \sum_{k=1}^{\frac{N}{2}-1} k \tag{3.62}$$

which can be further simplified to:

$$|x'(t)| = \frac{4\pi\sqrt{6}}{T\sqrt{N}} \frac{(N/2-1)(N/2-2)}{2} \tag{3.63}$$

$$\leq \frac{4\pi\sqrt{6}}{T\sqrt{N}} \frac{N^2}{8} = \frac{\sqrt{6}\pi N^{3/2}}{2T} \tag{3.64}$$

and therefore the maximum signal slope is bounded by

$$\max_t |x'(t)| \leq \frac{\sqrt{6}\pi N^{3/2}}{2T} \tag{3.65}$$

Since $x(t)$ is continuous and differentiable, the maximum value over the interval $t \in [pT/NL, (p+1)T/NL]$ given the corner values $x[p/L] = x(pT/NL)$ and $x[(p+1)/L] = x((p+1)T/NL)$ is (see Figure 3.8)

$$|x(t)| \leq \max \left\{ |x[p/L]| + \frac{\sqrt{6}\pi N^{3/2}}{2T} \frac{T}{2NL}, \right.$$
$$\left. |x[(p+1)/L]| + \frac{\sqrt{6}\pi N^{3/2}}{2T} \frac{T}{2NL} \right\} \tag{3.66}$$

$$= \max\{|x[p/L]|, |x[(p+1)/L]|\} + \frac{\sqrt{6}\pi\sqrt{N}}{4L} \tag{3.67}$$

$$\leq \max_n |x[n/L]| + \frac{\sqrt{6}\pi\sqrt{N}}{4L} \tag{3.68}$$

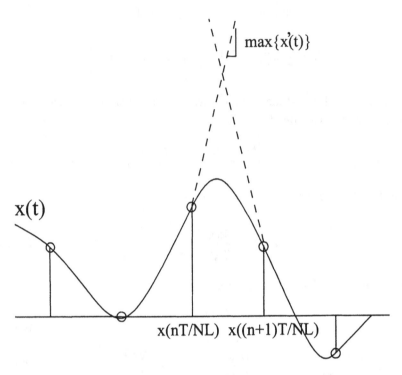

Figure 3.8: Upper bound on the maximum of $x(t)$ given $x[n/L]$.

Therefore, by oversampling at a rate $L = \kappa\sqrt{N}$ the maximum error in the amplitude estimate of $x(t)$ by using the discrete-time neighboring samples $x[p/L]$ and $x[(p+1)/L]$ is given by

$$\max ||x(t)| - x_{max}| \leq \frac{\pi\sqrt{6}}{4\kappa}, \quad t \in [pT/NL, (p+1)T/NL] \qquad (3.69)$$

where $x_{max} = \max\{|x[p/L]|, |x[(p+1)/L]|\}$. Therefore, increasing the number of subcarriers requires an increase in the oversampling rate necessary to guarantee that the continuous-time PAR is within a desired limit of the discrete-time PAR. In practice, the signal slope is well below the maximum just derived, i.e. $|x'(t)| \ll \frac{\pi\sqrt{6N}}{4L}$ with very high probability, and the error is much smaller than (3.69) most of the time. To derive error bounds that are more relevant in practice, we will derive the statistical distribution of the slope. Since the subcarrier values are uncorrelated, from (3.59), the mean and variance of $x'(t)$ are

$$E\{x'(t)\} = \frac{-2}{\sqrt{N}} \sum_{k=1}^{\frac{N}{2}-1} \left[E\{R_k\} \frac{2\pi k}{T} \sin(2\pi kt/T) \right.$$

$$\left. + E\{I_k\} \frac{2\pi k}{T} \cos(2\pi kt/T) \right] \tag{3.70}$$

$$= 0 \tag{3.71}$$

and

$$E\{|x'(t)|^2\} = \frac{4}{N} \sum_{k=1}^{\frac{N}{2}-1} \left[E\{|R_k|^2\} \left(\frac{2\pi k}{T}\right)^2 sin^2(2\pi kt/T) \right.$$

$$\left. + E\{|I_k|^2\} \left(\frac{2\pi k}{T}\right)^2 cos^2(2\pi kt/T) \right] \tag{3.72}$$

$$= \frac{4}{N} \sum_{k=1}^{\frac{N}{2}-1} \left[\frac{1}{2} \left(\frac{2\pi k}{T}\right)^2 sin^2(2\pi k/T) \right.$$

$$\left. + \frac{1}{2} \left(\frac{2\pi k}{T}\right)^2 cos^2(2\pi kt/T) \right] \tag{3.73}$$

$$= \frac{8\pi^2}{NT^2} \sum_{k=1}^{\frac{N}{2}-1} k^2 = \frac{8\pi^2}{NT^2} \frac{(N/2-1)(N/2)(N-1)}{6} \tag{3.74}$$

$$\leq \frac{8\pi^2}{NT^2} \frac{(N/2)^3}{3} = \frac{\pi^2 N^2}{3T^2} = \left(\frac{\pi N}{T\sqrt{3}}\right)^2 \tag{3.75}$$

The only random variables in (3.59) are R_k and I_k, which are typically discrete uniform bounded random variables. Therefore, from the central limit theorem, we should expect $x'(t)$ to be approximately Gaussian distributed for large values of N. Figure 3.9 compares the Gaussian probability density function (pdf) and the pdf of $x'(t)$ for $N = 128$ and 10^7 samples (notice the vertical log scale). Both pdfs match extremely well over the range where the Monte Carlo simulation is accurate ($pdf > 100/(num\ samples) = 10^{-5}$). Since for large N the derivative is approximately Gaussian distributed (with zero mean and standard deviation $\pi N/T\sqrt{3}$), the probability that the slope exceeds a given value

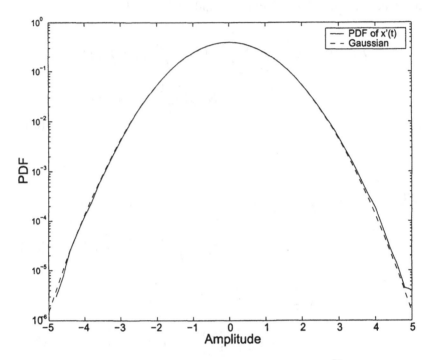

Figure 3.9: pdf of $x'(t)$ (normalized by $\pi N/T\sqrt{3}$) vs. a normalized Gaussian.

is given by

$$Prob\left\{x'(t) > \gamma\frac{\pi N}{T\sqrt{3}}\right\} = \frac{1}{\sqrt{2\pi}}\int_{\gamma}^{\infty} e^{-x^2/2}dx = Q(\gamma) \qquad (3.76)$$

Repeating the derivation in (3.60)-(3.68) we have the probabilistic upper bound

$$\max||x(t)| - x_{max}| \leq \gamma\frac{\pi N}{T\sqrt{3}}\frac{T}{2NL} \qquad (3.77)$$

$$= \gamma\frac{\pi}{2L\sqrt{3}} \qquad (3.78)$$

with probability $Q(\gamma)$. Again $x_{max} = \max\{|x[p/L]|, |x[(p+1)/L]|\}$ and $t \in \mathcal{T} = [pT/NL, (p+1)T/NL]$. For example, choosing the oversampling rate $L = 4$ and a 99.9% confidence ($Q(\gamma) = 10^{-3}$, $\gamma = 3$), the maximum difference between the continuous-time and discrete-time representation over the time interval $\mathcal{T}$ should be:

$$\max\,||x(t)| - x_{max}| \leq \gamma\frac{\pi}{2L\sqrt{3}} = 0.68 \qquad (3.79)$$

Therefore, from (3.78), the amount of oversampling necessary for high-ly reliable continuous-time PAR estimation does not increase with the number of tones. This statistical upper bound on the continuous-time PAR is still pessimistic as it assumes the statistical maximum slope is maintained over the full time interval between samples (positive over half the interval and negative over the other half – see dashed lines in Figure 3.8). Since $x'(t)$ is differentiable, the signal must smoothly switch slopes and will never achieve this maximum. Moreover this upper-bound can be lowered when $x[p/L] \neq x[(p+1)/L]$. Figures 3.7, 3.10 and 3.11 display some results on the effect of oversampling on the PAR CCDF for the ideal PAR reduction method $N = 256$. Figure 3.7 plots the CCDF of $PAR\{x^m[n/L]\}$ for $L = 2, 4$ given $PAR\{x^m[n]\} \leq 10$ dB. From (3.79), for $L = 1$ the maximum difference should be less than 2.7 with prob-ability $1 - Q(3) = .999$. Since $PAR\{x^m[n]\}$ has been reduced to 10 dB, which corresponds to a maximum amplitude of $3.16 = 10^{10/20}$ this im-plies that the statistical maximum is less than 15.3 dB. The probability of this maximum should be roughly $Prob\{x_{max} > 10dB\}Q(3) \approx 10^{-6}$. Furthermore, increasing the oversampling from $L = 2$ to $L = 4$ does not increase the PAR any more. In Figure 3.10 are plots of the CCDF of $PAR\{x^m[n/L]\}$ for $L = 4, 8$ given $PAR\{x^m[n/2]\} \leq 10$ dB. From (3.79), the maximum difference should be less than 1.35 and the the maximum is less than 13 dB, which is clearly an upper bound for the results shown. From the figure, the difference between the $L = 2$ and $L = 8$ PAR CCDF is only 1.5 dB and therefore $L = 2$ can be sufficient for PAR reduction techniques. Finally, in Figure 3.11 are plots of the effect of oversampling an $L = 4$ symbol. For this case, the PAR increase is small. The upper bound is 11.7 dB. The simulations in subsequent chapters verify that $L = 4$ is more than sufficient, and in many cases $L = 2$ gives good performance for our PAR reduction methods.

So far, ideal IFFT oversampling was assumed for $x^m[n/L]$. The over-sampling and continuous-time equivalents can be generated by other means, such as time-interpolation filters. For these cases, the upper bounds on the PAR increase depend on the discrete-time and continuous-time interpolation filters or pulse shaping filters. Fortunately, if these

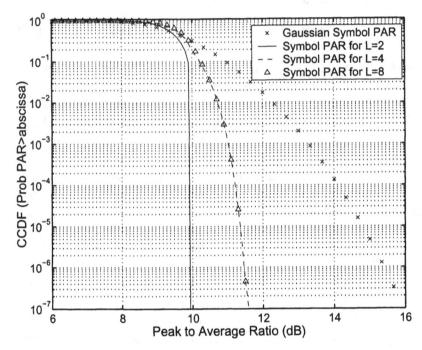

Figure 3.10: CCDF of $PAR\{x^m[n/L]\}$ for $L = 4, 8$ given
$PAR\{x^m[n/2]\} \leq 10\ dB$.

filters do not distort the the signal significantly, i.e. they are well behaved over the frequency band of the signal, most of these bounds are
good. A wide range of digital filters were tested to verify this, and several cases are included in the following chapters. Continuous-time signals
have been modeled by using very high oversampling, typically $L \geq 8$.

In Figure 3.12 the effect of oversampling is studied for an ideal PAR
reduction method with $L = 2$ and the PAR threshold set to 9.5 dB
for an ADSL application. In particular, in Figure 3.12 are plots of the
sample CCDFs at different points in the transmitter. The dashed line
is the CCDF of the oversampled IFFT output, $x^m[n/2]$ after ideal PAR
reduction. The continuous line is the sample CCDF after a 20-tap linear-
phase FIR High Pass Filter (HPF) with 35 dB attenuation in the stop-
band output (POTS band filter). The line with circles is the CCDF of
the continuous signal $x(t)$ after the DAC and an "analog" 40 tap FIR
LPF with 50 dB attenuation (to meet the PSD masks) simulated with
$L = 8$ oversampling. For this oversampled PAR reduction method the

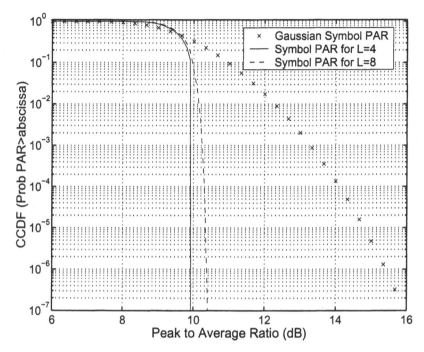

Figure 3.11: CCDF of $PAR\{x^m[n/8]\}$ given
$PAR\{x^m[n/4]\} \leq 10\ dB$.

PAR reduction loss after HPF and digital to continuous-time conversion
is less than 2 dB for a net PAR reduction of 4 dB at a sample clipping
rate of 10^{-8}. For a PAR reduction with 4× oversampling ($L = 4$), the
net PAR reduction is over 4.5 dB. Additional filters have been tested and
although the exact PAR reduction loss depends on the particular choice
of filters, these values still hold for a relatively large number of well-
designed filters. Subsequent chapters include some examples utilizing
filters provided by an ADSL chip developer.

5. DESCRIPTION OF MEMORYLESS NONLINEARITY

This section describes some nonlinear models that are commonly used
in the literature to represent commonly used nonlinear physical devices.
Section 6. studies the degradation effects of these nonlinearities on the
PSD and BER for multicarrier systems without PAR reduction. Theoretical limits on the achievable data rate (mutual information) for these

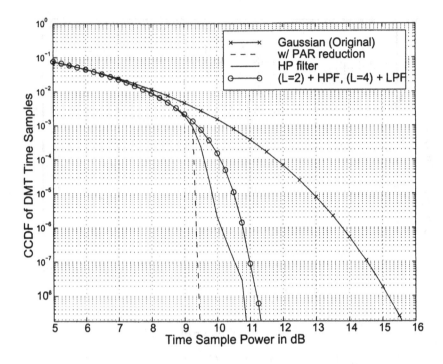

Figure 3.12: CCDF of $x[n/L]$ at different points of an
ADSL modem for an Ideal Over-Sampled
PAR reduction method.

common nonlinearities when the transmit distribution is not optimized
are studied in detail in Chapter 6.

With $g(\cdot)$ as the nonlinear function, the distorted, transmitted signal
is:

$$\mathbf{x}^g = g(\mathbf{x}) \qquad (3.80)$$

where $\mathbf{x}$ denotes either a discrete-time sequence of samples $x[n/L]$, or
the continuous-time signal $x(t)$.

To simplify the discussion further, this section will assume a memoryless nonlinearity that reduces (3.80) to a scalar equation. This simpli-
fying assumption is often made in the literature since many commonly
used nonlinear devices, such as limiters and high-power amplifiers, can
be accurately modeled as memoryless devices. Nevertheless, most of the

following concepts can be extended, with slight modifications, to non-linear devices with memory. If the memoryless nonlinear operation is performed on the discrete-time samples, the output can be expressed as:

$$x^g[n] = g(x[n]) \tag{3.81}$$

Similarly, if the nonlinear operation is performed on the continuous-time signal, the distorted signal can be expressed as:

$$x^g(t) = g(x(t)) \tag{3.82}$$

The following will consider only non-expansive nonlinearities with a maximum saturation value A. The non-expansive property can be written mathematically as

$$|g(x)| \leq |x|, \quad \forall x \tag{3.83}$$

Certainly, any nonlinear device characterized by a continuous function f and a maximum gain $\alpha > 0$ that satisfies $|f(x)| \leq \alpha|x|$, can be written as $f = \alpha g$ where g is non-expansive. Thus, using this fact, any nonlinear amplifying device can be decomposed into an ideal linear amplifier with gain α, followed by a non-expansive device described by g. Most practical nonlinear devices also exhibit a saturation property that can be expressed as

$$|g(x)| \leq A, \quad \forall x \tag{3.84}$$

There is a number of transceiver components with nonlinear behavior. The following lists a few models commonly used to describe the nonlinear characteristics of both DAC/ADC and Power Amplifiers. These models will be used extensively in this book to evaluate performance bounds on ideal transceivers and on the proposed PAR reduction structures. The most common sources of nonlinear behavior in the discrete-time domain are the DAC and the ADC. The main source of distortion for both DAC and ADC is the quantizer [Oppenheim and Schafer, 1989]. For quantization evaluation, it is convenient to represent the complex input signal in Cartesian coordinates as

$$x = \Re e\{x\} + j\Im m\{x\} \tag{3.85}$$

With this representation, the ideal input-output characteristic of a uniform quantizer is

$$Q(u) = \begin{cases} \Delta \, round(u/\Delta), & |u| \leq A \\ A \, sign(u), & |u| > A \end{cases} \quad (3.86)$$

where u represents $\Re e(x)$ or $\Im m(x)$. The quantization step-size is Δ and the saturation level is A. This is an example of a rounding quantizer with saturation [Oppenheim and Schafer, 1989]. Other closely related quantizers can be used (e.g. substituting truncation for rounding or natural overflow for saturation). They are well documented in the literature and are beyond the scope of this book.

Typically, most of the nonlinear behavior in the continuous-time domain is caused by the High Power Amplifiers (HPA). For most nonlinear HPA, it is convenient to represent the input signal in polar coordinates as

$$x = |x| \, e^{j \, \arg\{x\}} = \rho \, e^{j\phi} \quad (3.87)$$

Hence, the complex envelope of the output signal can be expressed by

$$g(x) = F[\rho] \, e^{j(\phi + \Phi[\rho])} \quad (3.88)$$

where $F[\rho]$ and $\Phi[\rho]$ represent, respectively, the AM/AM and AM/PM conversion characteristics of the memoryless nonlinearity. In particular, some of the most commonly used models for nonlinear amplifiers are [Rowe, 1982, Santella and Mazzenga, 1998]:

1. *Soft Limiter (SL)*

   The AM/AM and AM/PM nonlinear characteristics of a Soft Limiter (SL) can be written as

   $$F[\rho] = \begin{cases} \rho, & \rho \leq A \\ A, & \rho > A \end{cases} \quad (3.89)$$

   $$\Phi[\rho] = 0 \quad (3.90)$$

   Since the AM/PM component is zero the nonlinear characteristic of a SL can be rewritten as:

   $$g(x) = \begin{cases} x, & |\rho| \leq A \\ Ae^{j\phi}, & |\rho| > A \end{cases} \quad (3.91)$$

Although most physical components will not exhibit this piecewise linear behavior, the SL can be a good model if the nonlinear element is linearized by a suitable predistorter [D'Andrea et al., 1996, Andreoli et al., 1997, Jeon et al., 1997, Park and Powers, 1998].

2. *Solid-State Power Amplifiers (SSPA)*
The input-output relationship of many Solid-State Power Amplifiers (SSPA) can be modeled as:

$$F[\rho] = \frac{\rho}{[1 + (\frac{\rho}{A})^{2p}]^{1/2p}} \qquad (3.92)$$

$$\Phi[\rho] = 0 \qquad (3.93)$$

where the parameter p controls the smoothness of the transition from the linear region to the limiting or saturation region. When $p \to \infty$, the SSPA model approximates the SL characteristics.

3. *Traveling-wave Tube (TWT)*
The AM/AM and AM/PM characteristics for a TWT according to Saleh are [Saleh, 1981]

$$F[\rho] = \frac{\rho}{1 + (\rho/2A)^2} \qquad (3.94)$$

$$\Phi[\rho] = \frac{\pi}{3} \frac{\rho^2}{\rho^2 + 4A^2} \qquad (3.95)$$

All these models satisfy the non-expansive property and give a maximum output signal amplitude of A.

6.   EFFECT OF NONLINEARITIES ON SYSTEM PERFORMANCE

When the transceiver signal suffers from nonlinear distortion, the system experiences two main problems, namely, PSD degradation and BER increase. These problems are described in the following sections for the multicarrier case.

Since the multicarrier signals are modeled as Gaussian signals, the amount of distortion introduced by the memoryless nonlinearity depends only on the ratio $A^2/E\{|x|^2\}$, where A^2 is the maximum power output from the nonlinear device and $E\{|x|^2\}$ is the average energy of the input signal to the nonlinearity, we define a parameter called *ClipLevel* as:

$$ClipLevel = 10 \log_{10} \left(\frac{A^2}{E\{|x|^2\}} \right) [dB] \qquad (3.96)$$

Often, the term *Input Back-Off (IBO)* is often used in the literature to denote the same quantity. Similarly the term *Output Back-Off (OBO)* is used to denote

$$OBO = 10 \log_{10} \left(\frac{A^2}{E\{|g(x)|^2\}} \right) [dB] \qquad (3.97)$$

In general, the output power is smaller than the input power for any non-expansive nonlinearity. However, if the multicarrier signal operates in the linear region of the device most of the time, these two powers are very similar, and the $IBO \approx OBO$.

6.1   PSD DEGRADATION

The PSD plotted in Figure 3.4 only apply when the multicarrier signal does not suffer nonlinearities. If any of the signals described in Section 1. suffer nonlinear distortion, the output signal will suffer from inter-modulation distortion resulting in energy being generated at frequencies outside the allocated bandwidth, a phenomenon often denoted as *spectral regrowth*. The same signal in Figure 3.4 for $\beta = 16$ is plotted in Figure 3.13 for different values of *ClipLevel* or *IBO* and the SL nonlinearity. In many applications, the transceiver must share the spectra with users in adjacent channels, which may require large values of *ClipLevel* or filtering after the nonlinear device. Filtering after a HPA can be costly and in many cases the transmit power is lowered at the expense of increasing the decoded BER. For example if a -50 *dB* level of Adjacent Channel Interference (ACI) must be satisfied at a normalized frequency 1.2 relative to our signal, from Figure 3.4 11 *dB* of *IBO* are necessary.

6.2   BER INCREASE

This section studies the effect of nonlinearities on the DMT/OFDM symbol BER. First the degradation that results from treating the distortion as an AWGN term is quantified. This is the most common assumption in the literature, although it is quite inaccurate when one or more of the following conditions hold: the *ClipLevel* is high, the tone constellation size is large or the number of tones is small. New BER estimation formulas are proposed that are accurate over a larger range

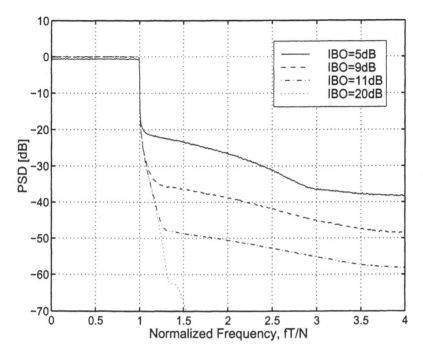

Figure 3.13: PSD for $x[n/4]$ with tapered window $w_T[n]$
followed by a limiter.

of modulation parameters. Since the distortion is a deterministic function of the transmit vectors, treating it as an additive independent term
is clearly pessimistic. In Chapter 6 a new ML receiver is formulated
that reduces the BER very significantly when the nonlinearity function
is known at the demodulator.

The mathematical derivation will use a discrete-time model of the
transceiver. If the only nonlinearity is in the discrete-time domain
(e.g. DAC), this analysis is exact. However, if the nonlinearity is present
in the continuous-time domain, this analysis will only be an approximation, but it is accurate if the oversampling of the discrete-time model
is increased to include the spectral regrowth from the nonlinearity. In
practice, since most of the energy from the nonlinearity will be in the
third-order inter-modulation distortion, 3× oversampling will often be
sufficient.

The output of the memoryless nonlinearity can be written as

$$x_\tau^g = g(x_\tau) = k^g x_\tau + d_\tau^{(\mathbf{X},g)} \tag{3.98}$$

Depending on the transmitter structure and the location of the nonlinearity, the signal x_τ can be replaced by any of the discrete-time or continuous-time signals defined in Section 1. The constant k^g is chosen to minimize the MSE term $E\{|g(x_\tau) - k^g x_\tau|^2\}$ [Friese, 1998]. Therefore, the sequence $d_\tau^{(\mathbf{X},g)} = g(x_\tau) - k^g x_\tau$ is the sequence that contains the minimum distortion energy and is also uncorrelated with x_τ. The notation $d_\tau^{(\mathbf{X},g)}$ is used to stress that d_τ is a function of the multicarrier MQAM/MPSK data vectors $\mathbf{X}$ and the nonlinear function $g(\cdot)$. If the nonlinearity is present after the digital filter for $z_F[n/L]$, or after the analog filter for $x_C(t)$ or $x_D(t)$, $d_\tau^{(\mathbf{X},g)}$ will also depend on pulse shaping filters $p[n/L]$ and/or $p(t)$. It can be shown [Friese, 1998] that for the SL and SSPA nonlinearities, $k^g \to 1$ for *ClipLevels* > 7 *dB*. Thus (3.98) can be approximated by

$$x_\tau^g = g(x_\tau) = x_\tau + d_\tau^{(\mathbf{X},g)} \tag{3.99}$$

In general, each distortion sample in the sequence $d_\tau^{(\mathbf{X},g)}$ can influence several received multicarrier vectors, but when the channel ISI is shorter than the CP, the memoryless distortion is limited in its effect to a single multicarrier symbol. For example, consider the case where the nonlinearity is performed on the sequence $x_{CP}[n/L]$ in (3.11). Since the distortion is a deterministic function of $x_{CP}[n/L]$, then the distortion term $d^{(\mathbf{X},g)}[n/L]$ and the distorted transmit sequence $x^g[n/L]$ in (3.99) will also satisfy the cyclic-prefix property. Therefore, the equivalent transmit QAM vector for the *m-th* symbol $\mathbf{X}_L^m$, including the nonlinear effect can be computed from the DFT of (3.99) for the case $x_\tau^g = x^g[n/L]$ over the discrete-time interval $(m(N+\nu)L, NL-1+m(N+\nu)L))$, i.e.

$$DFT(x_\tau^{m,g}) = \mathbf{X}_L^{m,g} = \mathbf{X}_L^m + \mathbf{D}_L^{(\mathbf{X}^m,g)} = \tag{3.100}$$

$$= \begin{cases} X_k^m + D_{L,k}^{(\mathbf{X}^m,g)}, & k = 0,\dots,\frac{N}{2}-1 \\ X_{k-N(L-1)}^m + D_{L,k}^{(\mathbf{X}^m,g)}, & k = NL-\frac{N}{2},\dots,NL-1 \\ D_{L,k}^{(\mathbf{X}^m,g)}, & k = \frac{N}{2},\dots,NL-\frac{N}{2} \end{cases} \tag{3.101}$$

The distortion terms $D_{L,k}^{(\mathbf{X}^m,g)}$, $k = \frac{N}{2},\dots,NL-\frac{N}{2}$ represent the out-of-band distortion and can be minimized with clip windowing [Pauli and

Kuchenbecker, 1997, van Nee and de Wild, 1998, Pauli and Kuchenbecker, 1998] or by using smoother nonlinearities. If the nonlinearity is
present on Nyquist rate samples (e.g. a saturating DAC with $L = 1$),
then, these extra terms will not be present. For the case of nonlinear
distortion on $x_C(t)$ or $z_T[n/L]$, (3.101) will also apply since the cyclic
prefix structure is preserved, but will only be an approximation if digital $(p[n/L])$ or analog $(p(t))$ filtering precedes the nonlinearity. The
decomposition of the sequential signal transmission into independen-
t symbols allows working with each multicarrier vector independently,
and the symbol index m may be dropped to simplify the notation. For
each symbol, the distortion term $d_T^{(\mathbf{X},g)}$ is a deterministic function of the
random vector $\mathbf{x}$ with average power given by:

$$E\{|d_T^{(\mathbf{X},g)}|^2\} = \int_{-\infty}^{\infty} (x - x^g)^2 p_x(x) dx \qquad (3.102)$$

For the SL nonlinearity and using the Gaussian approximation for
$x^m[n/L]$ the variance of the distortion has a closed form [Mestdagh et al.,
1994]:

$$\sigma_d^2 = E\{|d_T^{(\mathbf{X},g)}|^2\} \qquad (3.103)$$

$$= \frac{2}{\sqrt{2\pi}\sigma_x} \int_A^{\infty} (x - A)^2 e^{-x^2/2\sigma_x^2} dx \qquad (3.104)$$

$$= \sigma_x^2 \left(-\sqrt{\frac{2}{\pi}} \mu e^{-\mu^2/2} + (1 + \mu^2) erfc(\mu/\sqrt{2}) \right) \qquad (3.105)$$

where $\sigma_x^2 = E\{|x|^2\}$ is the power of the desired signal, and $\mu = A/\sigma_x$ is
the square root of *ClipLevel* on a linear scale.

The DFT of the distortion, $D_{L,k}^{(\mathbf{X}^m,g)}$ is also a deterministic function
of the random vector $\mathbf{x}$. If the symbol size is large, and the number of
nonzero distortion terms is also large (i.e. there is a lot of distortion),
from the Central Limit Theorem, the DFT of the distortion will result
in approximately Gaussian random variables. For the critically sampled
case, $L = 1$, with Gaussian distributed subsymbols, the symbol samples are i.i.d. (Gaussian) and therefore the distortion samples are also
i.i.d. (but not Gaussian). Thus, the distortion frequency samples are
also i.i.d. and the Signal to Noise plus Distortion Ratio (SNDR) is given

by

$$SNDR = \frac{|H_k|^2 \sigma_{X,k}^2}{|H_k|^2 \sigma_{D,k}^2 + \sigma_{N,k}^2} \qquad (3.106)$$

where $\sigma_{D,k}^2 = \sigma_d^2$ for the i.i.d. case, $|H_k|^2$ is the channel gain and $\sigma_{N,k}^2$ is the receiver noise variance at subsymbol k. Assuming the distortion is an independent AWGN term the Subsymbol Error Rate (SER) for MQAM subsymbols can be approximated as:

$$SER \approx 4 \left(1 - \frac{1}{\sqrt{M}}\right) Q \left(\sqrt{\frac{3SNDR}{M-1}}\right) \qquad (3.107)$$

This approximation is very accurate for large multicarrier symbols (large N) with few bits per tone (e.g. 4QAM and 16QAM). For example the authors in [O'Neill and Lopes, 1994] compare analytical and simulated results for symbol size $N = 256$ and 4QAM and 16QAM constellations. The match is very good for the 4QAM case but the Gaussian approximation is slightly optimistic for the 16QAM constellation.

Figure 3.14 compares analytical and simulated results for a multicarrier symbol size of $N = 512$ and 64QAM constellations and the SL nonlinearity. For this case, we set $ClipLevel = 8\ dB$ which corresponds to a Signal to Distortion Ratio of $SDR = 27\ dB$ as computed from (3.105). In this figure, the Equivalent SNR, which is defined as $\frac{3SNR}{M-1}$ is varied over the range 7-16 dB. For this case, the simulated results and the analytical results based on the Gaussian approximation in (3.107) differ about an order of magnitude for $SER < 10^{-3}$. The curve denoted Gaussian Approx B, was generated by (3.113). This novel approximation is closer to the simulated curve and is describer later in this section.

In Figure 3.15, the constellation size is increased to 1024QAM and the linear range is increased to $ClipLevel = 11\ dB$. Since the constellation size has increased, the analytical results are much less reliable. For low SER the difference is up to three orders of magnitude. The remaining part of this section is devoted to a new analytical method, denoted Gaussian Approximation B, which matches the measured SER more closely.

The main reason the described approximation fails is the impulsive nature of the distortion. When the $ClipLevel$ is increased the number of distorted samples is small and the amount of distortion can vary significantly on a symbol-by-symbol basis. Returning to the SL nonlinearity,

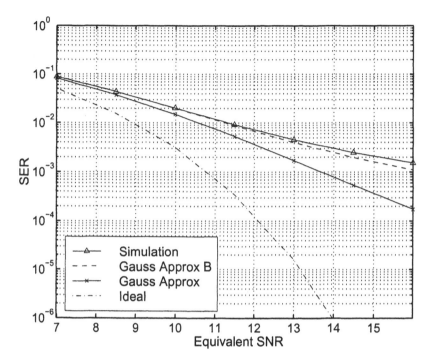

Figure 3.14: Analytical and simulated SER for $N = 512$ and 64QAM for the $SL(8\ dB)$ nonlinearity.

Figure 3.16 shows the pdf of the symbol samples, before and after the nonlinearity and the pdf of the distortion samples. Assuming real symbols, the samples of $\mathbf{x}$ are less than A with probability $1 - 2Q(\mu)$, and therefore the pdf of the distortion has a Dirac delta at zero amplitude.

Thus the distortion energy can be split into two different cases whether or not a sample experiences distortion.

$$\sigma^2_{d|clip} = E\{(x - g(x))^2 | |x| > A\} \tag{3.108}$$

$$= \frac{1}{Q(\mu)\sqrt{2\pi}\sigma_x} \int_A^\infty (x - g(x))^2 e^{\frac{-x^2}{2\sigma_x^2}}\, dx \tag{3.109}$$

$$\sigma^2_{d|no\ clip} = E\{(x - g(x))^2 | |x| < A\} = 0 \tag{3.110}$$

Since the probability of a distorted sample is $2Q(\mu)$ and the samples are assumed i.i.d., the number of distorted samples per symbol is a

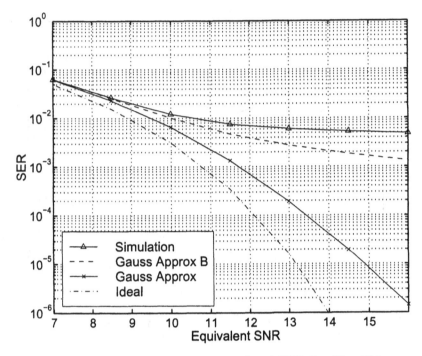

Figure 3.15: Analytical and simulated SER for $N = 512$
and 1024QAM for the SL (11 dB)
nonlinearity.

Binomial($N,2Q(\mu)$) that is

$$p_l = \binom{N}{l}(1 - 2Q(\mu))^l(2Q(\mu))^{N-l} \tag{3.111}$$

where p_l is the probability that l samples are clipped. Thus, SNDR at
tone k for l clipped samples is

$$SNDR(l) = \frac{|H_k|^2\sigma_{X,k}^2}{\frac{l}{N}|H_k|^2\sigma_{d|clip}^2 + \sigma_{N,k}^2} \tag{3.112}$$

In (3.112) the clip energy is divided by N since the distortion will be
divided among N tones. Assuming the DFT of the distortion is still

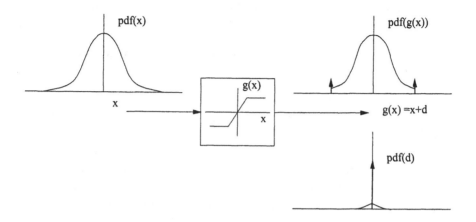

Figure 3.16: Diagram for the pdf of x, $g(x)$ and $x - g(x)$.

approximately Gaussian, the SER can be easily computed as

$$SER \approx 4 \left(1 - \frac{1}{\sqrt{M}}\right) \sum_{l=0}^{N} p_l Q \left(\sqrt{\frac{3SNDR(l)}{M-1}}\right) \qquad (3.113)$$

This analytical approximation is denoted Gauss Approx B in Figures 3.14 and 3.15 and is a better match to the simulated results. This formulation is much more accurate because it takes into account the impulsive nature of the distortion but is not exact because it assumes that when clipping occurs i.e. $|x| > A$, the distortion is Gaussian.

7.   LIMITS FOR DISTORTIONLESS PAR REDUCTION

Let's consider transmission over the discrete-time AWGN channel

$$y = x + n \qquad (3.114)$$

where x is the desired signal of average power $\sigma_x^2 = E\{|x|^2\}$, y is the received signal and n is the additive Gaussian noise of power $\sigma_n^2 = E\{|n|^2\}$, which is independent of x. Shannon proved [Shannon, 1948a, Shannon, 1948b] that the distribution on x that maximizes the mutual information between x and y, denoted here as $I(x; y)$, is the zero mean Gaussian distribution with variance σ_x^2. This maximum mutual information is called the Capacity of the channel and is given by

$$
\begin{aligned}
C(\sigma_x^2) &= sup_{p_x} I(x; y) = sup_{p_x} I(x; x + n) & (3.115) \\
&= sup_{p_x} h(x + n) - h(n) & (3.116) \\
&= \frac{1}{2} \log_2(1 + \frac{\sigma_x^2}{\sigma_n^2}) & (3.117)
\end{aligned}
$$

where $p_x = \mathcal{N}(0, \sigma_x^2)$ is the probability density function for x. In the formulae above, $h(z)$ denotes the differential entropy of a random variable z and is defined to be [Cover and Thomas, 1991]

$$h(z) = -\int p(z) \log_2 p(z) dz \qquad (3.118)$$

where $p(z)$ is the probability density function for z. $C(\sigma_x^2)$ is the Capacity of the Average-Power-Limited channel since the only constraint on x is on its variance. By taking the DFT of a Capacity achieving codebook for x [Cover and Thomas, 1991], we can easily construct an equivalent multicarrier codebook. From the properties of the DFT described in Section 3., this multicarrier codebook also has i.i.d. Gaussian components X_k. Since the codewords of x must be arbitrarily long, the multicarrier codewords will also be arbitrarily long.

Unfortunately, the Gaussian distribution is not bounded and therefore has infinite PAR. Adding an additional constraint on the distribution x, such as a peak power constraint $|x|^2 \leq P_{peak}$ will lead to a reduction of maximum data rate but will also lead to finite PAR that can be implemented. Under peak power and average power constraints, the transmit distribution that achieves capacity is discrete and has been derived in

[Smith, 1969] for the real case and in [Shamai (Shitz) and Bar-David, 1995] for the complex case. These capacity achieving discrete distributions do not have closed form expressions and must be solved numerically for each value of σ_x^2/σ_n^2. Fortunately, under moderate to high SNR conditions, i.e. $\sigma_x^2 \gg \sigma_n^2$ the entropy power inequality provides a tight lower bound. Thus the Capacity of a Peak-and-Average-Power-Limited channel can be lower bounded by

$$
\begin{aligned}
C(\sigma_x^2, P_{peak}) &= sup_{p_x} I(x; y) = sup_{p_x} I(x; x + n) & (3.119) \\
&= sup_{p_x} h(x + n) - h(n) & (3.120) \\
&\geq sup_{p_x} \log_2(2^{2h(x)} + 2^{2h(n)}) - h(n) & (3.121) \\
&= sup_{p_x} \log_2(2^{2h(x)} + 2\pi e \sigma_n^2) - \frac{1}{2} \log_2(2\pi e \sigma_n^2) & (3.122)
\end{aligned}
$$

The maximum of (3.122) is achieved when the pdf of x is a truncated Gaussian, $p_x = a e^{-x^2/2b^2}, |x| < \sqrt{P_{peak}}$, where the constants a and b are chosen such that $E\{|x|^2\} = \sigma_x^2$ and the pdf integrates to 1. The resulting differential entropy of x is [Shamai (Shitz) and Dembo, 1994]

$$
h(x) = \frac{1}{2} \log_2(2\pi e b^2) + \log_2(erf(A/b\sqrt{2})) - \frac{A \log_2(e) exp(-A^2/2b^2)}{b\sqrt{2\pi} \, erf(A/b\sqrt{2}))}
$$
(3.123)

where $A = \sqrt{P_{peak}}$, $erf(x) = 1/\sqrt{\pi} \int_0^{x/\sqrt{2}} exp(-u^2) du$ and b is computed from

$$
\sigma_x^2 = b^2 - \frac{2bA \, exp(-A^2/2b^2)}{\sqrt{2\pi} erf(A/b\sqrt{2})}
$$
(3.124)

Figure 3.17 shows the capacity for the Peak-and-Average-Power-Limited transmitter relative to the Average-Power-Limited transmitter for different values of PAR. These lower bounds imply that PAR values of 9 *dB* are possible with very small penalty in achievable data rate.

The corresponding low PAR multicarrier symbols are the IDFT of the codebooks generated from truncated Gaussian (or discrete) i.i.d. random vectors of arbitrary length. In this case, the elements in the multicarrier symbols, X_k will no longer be independent.

Unfortunately, the Peak-and-Average-Power-Limited capacity lower bound described in this section only applies in the context of infinite

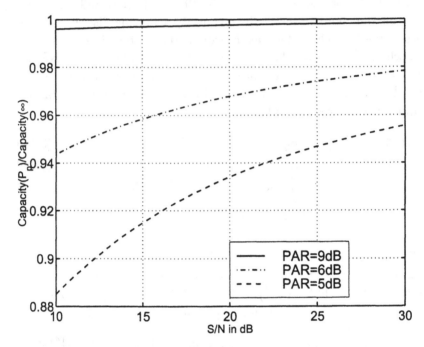

Figure 3.17: Relative Capacity of Peak-Power-Limited
AWGN channel.

complexity and delay. Moreover, these capacity bounds have been derived for i.i.d. discrete-time samples (i.e. critically sampled signals) and
do not include the continuous-time PAR increases. Therefore this bound
should be considered only as theoretical limit of performance for comparing practical PAR reduction schemes applied to critically sampled
multicarrier symbols ($L = 1$). In [Shannon, 1948a, Shannon, 1948b],
Shannon derived a peak-power capacity lower bound for ideal bandpass
continuous-time signals

$$C(P_{peak}) \geq \frac{1}{2} \log_2 \left(\frac{2}{\pi e^3} \frac{P_{peak}}{\sigma_n^2} \right) \qquad (3.125)$$

For high SNR, the additive "1" in (3.117) can be neglected. Thus,
comparing (3.117) and (3.125), this Peak-Power-Limited capacity lower

bound is equivalent to a $\frac{2}{\pi e^3}$ or 15 dB power reduction relative Average-Power-Limited capacity. For the quasi-bandpass[4] windowed multicarrier signals defined in Section 1., the practical PAR reduction methods described in the following chapters can achieve PAR below 11 dB for the continuous-time signal.

8.   TECHNIQUES FOR PAR REDUCTION

As we have shown in Section 7., it is theoretically possible to reduce the PAR significantly with low reductions in maximum achievable data rate. This does not imply that there exists any practical scheme that can achieve these PAR and data rates with finite complexity (i.e. non exponential) and finite delay. We will now proceed to describe some of the most interesting techniques proposed to date for PAR reduction. The use of (orthogonal) multicarrier signals for data transmission is a relatively recent phenomenon and therefore techniques for PAR reduction were not proposed until the mid 90's [Jones et al., 1994]. On the other hand, equally spaced and equal power multicarrier signals have been used earlier for linear system frequency response measurement, as a method to generate signals with perfect low-pass or bandpass spectra. In this case, the target is to excite all possible frequency modes with as much energy as possible, to increase the accuracy of the measurement, without large instantaneous time domain values that can damage the device under test or produce undesired nonlinear effects. For this application, we only need one signal with the lowest possible PAR, and therefore no information can be transmitted. Since all the carriers must have the same amplitude, the only optimization parameter is the phase of each tone. Unfortunately, the problem of minimizing the PAR over the phase vector is a nonlinear problem, and no choice yields the global minimum PAR. However, two general techniques have been proposed for generating "good" phase vectors,

- Structured phase vectors. Some structured phase vectors which have been shown to generate low PAR multicarrier signals are the Newman Phases [Greenstein and Fitzgerald, 1981, Boyd, 1986], the Shapiro and Rudin Phases [Boyd, 1986], the Golay complementary sequences [Golay, 1961, Popovic, 1991] and the Narahashi phases [Narahashi and Nojima, 1994].

---

[4]For this case we use the energy bandwidth, i.e. the bandwidth which includes a very high percent of the signal energy.

- Optimized phase vectors. A number of nonlinear iterative algorithms have been proposed to achieve low PAR signals [Van der Ouderaa et al., 1988, Narahashi et al., 1995, Friese, 1997b]. These techniques can achieve lower PAR values, but they usually involve a large number of computations. Fortunately this only needs to be done once for each vector length and we can store these phase vectors.

When using multicarrier signals for transmission, the transmitter must have a large number of possible transmit symbols with low PAR to maximize the number of bits received per symbol. This is upper-bounded by $\log_2(|symbols|)$, where $|symbols|$ denotes the cardinality of the transmit symbol set. The techniques listed so far have provided us with lower-bounds on the achievable PAR, but more importantly, have provided us with some insight to their data transmission extensions as will become apparent in the following discussion.

The literature on PAR-reduction for multicarrier data transmission is relatively recent, but given the importance of this problem, it has been extremely prolific[5]. It is difficult to classify all these efforts into a simple outline or to summarize all these techniques in a few of pages without leaving a few interesting methods out of the picture. The following is probably the most commonly accepted classification. At a higher level this classification divides the methods into PAR-reduction with Distortion and Distortionless PAR-reduction. In the first set of techniques, the transmitter is not designed for the maximum PAR range and the transmit symbols are distorted. These methods degrade the decoded BER and are described in more detail in Section 8.1 The latter methods reduce the symbol PAR prior to the nonlinear device without increasing the BER. These methods are described in Section 8.2 and typically achieve lower PAR at the expense of a reduced data rate.

8.1   PAR REDUCTION WITH DISTORTION

Probably the simplest way to reduce the PAR of the DMT/OFDM transmit sequence is to clip the signal at the transmitter to the desired level. This operation can be implemented on the discrete samples prior to the DAC or by designing the DAC and/or amplifiers with saturation levels that are lower than the signal dynamic range. This approach is widely used although it degrades the received SER and increases the out-of-band radiation. There has been a number of publications that

---

[5]To date we are aware of at least 60 papers on this topic.

quantify the multicarrier signal degradation due to nonlinearities. Some of the early work on this topic was done in [Mestdagh et al., 1994, Gross and Veeneman, 1994] for real (baseband) multicarrier signals and in [O'Neill and Lopes, 1994, O'Neill and Lopes, 1995] for complex (pass-band) multicarrier signals. All of these papers assume the nonlinearity is an ideal soft limiter and compute the SNR degradation and the PSD of the clipped/limited signal. The out-of-band radiation can be reduced by filtering after the nonlinearity. Unfortunately filtering after a high power amplifier can be very costly. In practice, the clipping and filtering is done prior to the HPA but for this case the PAR can significantly increase after the filter [Li and Cimini, Jr., 1997]. An alternative method is clip windowing [Pauli and Kuchenbecker, 1997, van Nee and de Wild, 1998, Pauli and Kuchenbecker, 1998], where the clipping operation is smoothed with a time window. This reduces the out-of-band radiation but increases the BER. For some applications, the channel exhibits a known structure, and the clip/distortion noise, which is approximately white over the band of interest can be shaped to reduce the BER degradation. For example in XDSL applications, the channels are typically lowpass, and the high-bit constellations that are less robust against distortion, are located at the lower frequencies. By highpass filtering the distortion, we can improve the BER of our system. This is called shaped clipping [Chow et al., 1997a, Chow et al., 1997b].

Although the latter methods can reduce the spectral regrowth problem, unfortunately none of these methods mitigates the BER increase for a general channel. This increase in BER can be corrected with error correcting codes, but at the expense of a reduction in the effective data rate, an increase in the complexity at both the transmitter and receiver ends and an increase in the overall delay or latency of the system. In Chapter 6, we describe a novel method for correcting for added clipping noise with negligible increase in BER at the expense of additional complexity at the receiver.

8.2   DISTORTIONLESS PAR REDUCTION

These methods reduce the symbol PAR prior to the nonlinear device without increasing the BER. As described in Section 7., if the subsymbols X_k are chosen from the optimal i.i.d. Gaussian distribution, this PAR reduction must come at the expense of a data rate reduction. Since most practical multicarrier modulators use simple suboptimal uniform QAM mapping, it is possible to reduce PAR without any data rate loss.

A novel method based on this idea is described in Chapter 5. Most of the Distortionless PAR-reduction techniques to date can be classified into three main groups: *Coding, Discrete Parameter Optimization* (or Multiple Signal Representation) and *Continuous Parameter Optimization*. Unfortunately, most of the PAR-reduction methods in the latter two groups, as described in their corresponding references, have only been evaluated at the output of the critically sampled IDFT or $L = 1$ case. Although most of the methods can be extended to the oversampled and continuous-time signals, the quoted values of achievable PAR reduction, the complexity estimates and the algorithms described for these methods refer to these IDFT sample points only and are therefore optimistic.

8.2.1   CODING

As described above, there is a number of structured phase vectors that generate low PAR multicarrier symbols $x^m(t)$. To maximize the number of bits transmitted per multicarrier symbol, there must be a large number of phase vectors or codewords, ideally proportional to 2^N. So far, no set of this cardinality has been found but there are some promising initial results. In [Popovic, 1991], the author showed that Golay complementary sequences generate multicarrier symbols with very low PAR (3 *dB* complex baseband or 6 *dB* passband). These methods have been further studied in [van Nee, 1996, Davis and Jedwab, 1997]. Unfortunately, these codes suffer two big limitations: they can only be applied to MPSK subsymbols and the number of bits per multicarrier symbol is proportional to $\log_2 N$, i.e. the number of possible multicarrier symbols is $2^{K \log_2 N} = N^K \ll 2^N$. Since the code rate is proportional to $(\log_2 N)/N$, these methods are not practical for $N > 32$. In [Paterson, 1998], the author showed that the set of codewords can be increased at the expense of increasing the PAR by another factor of 2. The resulting symbol set is larger, but still remains proportional to $\log_2 N$. An advantage of the Coding methods is that they provide error correction capabilities.

Since all coding techniques to date lead to a very small set of possible multicarrier symbols, exhaustive searches for low PAR symbols have been performed [Jones et al., 1994, Wilkinson and Jones, 1995, Jones and Wilkinson, 1996, Shepherd et al., 1998]. These methods yield large reductions of PAR at the expense of minimal data rate loss. Unfortunately the necessary complexity to search and store these codewords is exponential in the number of tones and thus is not practical for $N > 16$. In particular the number of symbols to search over is M^N, where M is the constellation size. For example, using the smallest QAM constellation, i.e. 4QAM, the number of stored multicarrier symbols for $N = 32$ is on the order of 10^{18} symbols.

8.2.2   DISCRETE PARAMETER OPTIMIZATION

We use the term Discrete parameter optimization to denote all PAR reduction methods that can be formulated as

$$\min_{s} PAR\{\mathcal{T}^s(\mathbf{X}_L^m)\}, \quad s = 1, \dots, S \tag{3.126}$$

or equivalently as

$$\min_{s} PAR\{\mathcal{T}^s(x^m[n/L])\}, \quad s = 1, \dots, S \tag{3.127}$$

where $\mathcal{T}^s$ represents a reversible transformation of the multicarrier frequency-domain vector $\mathbf{X}_L^m$ or the time-domain sequence $x^m[n/L]$ and S is the number of possible transformations of the symbol. For these transformations to be practical the mapping must be reversible and the reliability of the data should be maintained, i.e. the SER of data vector $\mathbf{X}_L^m$ must be preserved within reasonable limits. For most of these methods the transformation chosen at the transmitter must be communicated to the receiver, and therefore these techniques incur an additional $\log_2(S)$ bits of overhead data that must be sent as side information. Probably the simplest technique based on this structure was proposed in [Wulich, 1996, Chow et al., 1997a, Chow et al., 1997b], where $\mathcal{T}^s$ is just a scaling operation:

$$\mathcal{T}^s(x^m[n/L]) = \alpha_s x^m[n/L] \tag{3.128}$$

where the set of scalars range over $0 < \alpha_s \leq 1$. Basically these techniques attenuate the symbol when the peak power is above some thresholds. This method is very simple, but suffers from increased SER since the scaling factor reduces the SNR at the receiver.

To avoid SER degradation, a number of researchers have independently proposed transformations based on phase shifts of the data subsymbols, i.e.

$$\mathcal{T}^s(\mathbf{X}^m) = [\mathcal{T}_0^s(X_0^m) \dots \mathcal{T}_k^s(X_k^m) \dots \mathcal{T}_{N-1}^s(X_{N-1}^m)] \quad (3.129)$$
$$= [e^{j\phi_0^s}X_0^m \dots e^{j\phi_k^s}X_k^m \dots e^{j\phi_{N-1}^s}X_{N-1}^m] \quad (3.130)$$

Some authors have proposed pseudo-random phase terms [Mestdagh and Spruyt, 1996, Van Eetvelt et al., 1996, Bäuml et al., 1996, Müller et al., 1997, Djokovic, 1997, Müller and Huber, 1997c, Müller and Huber, 1997a]. If the phase terms are chosen pseudo-randomly and the subcarrier constellations are small or the symbols are long[6], the resulting time-domain symbols will have PAR that are approximately independent. The limitation of these techniques is that for each phase transformation, the DFT must be recalculated, resulting in a worst case increase of transmitter complexity proportional to S. To avoid this large complexity some authors have proposed structured phase transformations [Müller and Huber, 1997b, Verbin, 1997, Müller and Huber, 1997c, Müller and Huber, 1997a, Zekri and Van Biesen, 1999]. These structured phase transformations exploit mathematical properties of the DFT to yield savings in recomputing the IFFT for each transformation of $\mathcal{T}^s$. These reductions in complexity come at the expense of reduced PAR reduction, since the transformed symbols are no longer independent from the original symbol. Typically a larger set of transformations are necessary to achieve the same PAR reduction and the complexity savings can vanish.

Several other interesting PAR reduction methods that conform to (3.127) or (3.126) have been proposed in the literature (e.g. [Henkel and Wagner, 1997, Zekri et al., 1998]) but their description is beyond the scope of this book. Finally, the Tone Injection method that is described in Chapter 5 is also a Discrete Parameter Optimization method.

8.2.3   CONTINUOUS PARAMETER OPTIMIZATION

We use the term Discrete Parameter Optimization to denote all PAR reduction methods that can be formulated as

$$\min_{\gamma_0, \gamma_1, \dots, \gamma_G} PAR\{\bar{x}^m[n/L]\}, \quad (3.131)$$

---

[6]If N is small, for large constellation sizes the power of a given multicarrier symbol can differ significantly from the average transmission power and the peak power of the tone rotated symbols will be partially correlated.

where

$$\bar{x}^m[n/L] = \mathcal{F}\left(\mathbf{X}^m, \gamma_0, \gamma_1, \ldots, \gamma_G\right) \qquad (3.132)$$

and $\gamma_0, \gamma_1, \ldots, \gamma_G$ are real valued parameters. That is, the transmitted symbol is a function of the original symbol and a set of design parameters which must be optimized. Two methods are included in this class: the Tone Reservation method that is described in Chapter 4, and the method described in [Friese, 1996, Friese, 1997a]. In [Friese, 1996, Friese, 1997a] the author proposed a PAR reduction method based on optimizing the real valued phase vector $\{\phi_0, \phi_1, \ldots, \phi_{\frac{N}{N_g}-1}\}$ in:

$$
\begin{aligned}
\bar{x}^m[n/L] \;=\; & \frac{1}{\sqrt{N}} \Bigg(e^{\phi_0} \sum_{k=-\frac{N}{2}}^{-\frac{N}{2}+N_g-1} X_k^m e^{j2\pi kn/NL} \\
& + e^{\phi_1} \sum_{k=-\frac{N}{2}+N_g}^{-\frac{N}{2}+2N_g-1} X_k^m e^{j2\pi kn/NL} \;+\; \cdots \\
& + e^{\phi_{\frac{N}{N_g}-1}} \sum_{k=\frac{N}{2}-N_g}^{N/2-1} X_k^m e^{j2\pi kn/NL} \Bigg) w[n/L] \qquad (3.133)
\end{aligned}
$$

This method divides the data symbol in groups of N_g tones and rotates each group by a constant phasor to minimize the PAR. Since each group of tones is rotated by a real number, all the constellations points are rotated by a quantity unknown at the receiver and only differential detection is possible. Moreover the leading tones in each of the N/N_g groups cannot carry any phase information. This method can yield large reductions in PAR but the PAR is a nonlinear function of the optimization parameters $\{\phi_0, \phi_1, \ldots, \phi_{\frac{N}{N_g}-1}\}$. Therefore, the iterative optimization algorithm proposed by the author are rather complex and may require up to 60 oversampled FFT per transmitted multicarrier symbol to produce good results.

9.   NEW PAR REDUCTION STRUCTURES

Chapters 4, 5 and 6 describe our three new methods for PAR reduction. All these structures can be formulated as:

$$\bar{x}^m[n/L] = x^m[n/L] + c^m[n/L] \qquad (3.134)$$

$$= \frac{1}{\sqrt{N}} \sum_{k=-\frac{N}{2}}^{\frac{N}{2}-1} (X_k^m + C_k^m) e^{j2\pi kn/NL}, \qquad (3.135)$$

where the frequency vector $\mathbf{C}^m = [C_0^m \dots C_{N-1}^m]$ or equivalently, the time domain sequence $\mathbf{c}^m = c^m[n/L]$ are the PAR reduction signals. The *bar* notation $\bar{x}[n/L]$, $\bar{z}[n/L]$, $\dots$, $\bar{x}(t)$ is used to denote PAR reduced signals. Given this common formulation, all these methods fall under a category that we denote *Additive model for PAR reduction*, and can be illustrated by the structure in Figure 3.18. For these structures to be effective methods for PAR reduction, they must satisfy most of the following list of desirable properties:

1. The PAR reduction signal must achieve significant reductions of the PAR of the combined signal $x^m[n/L] + c^m[n/L]$. The PAR of this combined signal is defined as:

$$PAR\{x^m[n/L] + c^m[n/L]\} = \frac{\max_n |x^m[n/L] + c^m[n/L]|^2}{E\{|x^m[n/L]|^2\}} \qquad (3.136)$$

   Notice that the energy of $c^m[n/L]$ was not included in the denominator of this PAR definition. This definition avoids misleading PAR reductions due to potential increases in the combined signal average power and therefore all the PAR reduction must come from peak power reduction.

2. The receiver must decode $\mathbf{X}^m$ efficiently from the combined vector $\mathbf{X}^m + \mathbf{C}^m$ without degrading the performance of the modem. Preferably the transmitter should not need to communicate side information for canceling $\mathbf{C}^m$ for each symbol.

3. Since the transmitter must compute the PAR reduction signals for many transmit symbols, efficient algorithms must exist.

4. These additive signals should not reduce the data throughput significantly.

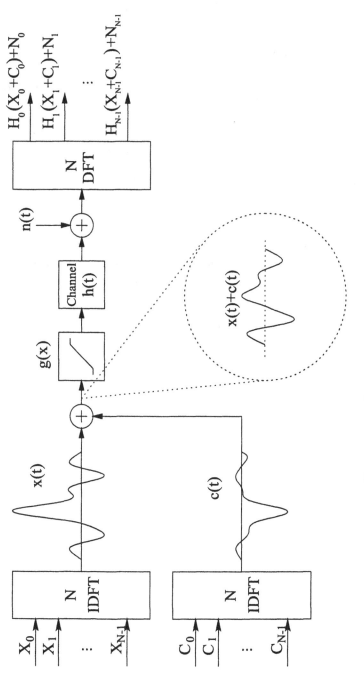

Figure 3.18: Additive model for PAR reduction.

5. These PAR reduction structures should not prevent using coding techniques (block codes, convolutional codes, turbo codes, etc,...)

Some additional desirable properties include the option for constraints on the combined transmit power, $E\{|\mathbf{X}+\mathbf{C}|^2\}$, or on each of the individual carriers, e.g., $|X_k^m + C_k^m|^2$.

All these PAR reduction methods are described in detail in the following chapters but are listed here for completeness:

- **Tone Reservation**: For which the data signal $\mathbf{X}^m$ and the correction signal $\mathbf{C}^m$, occupy Disjoint Frequency Bins, i.e. $X_k^m C_k^m = 0$.

- **Tone Injection**: Which choose $\mathbf{C}^m$ such that $\mathbf{X}^m = modulo_D(\mathbf{X}^m + \mathbf{C}^m)$, where $modulo_D$ is a vector operation on all the tones i.e. $X_k^m = mod_{D_k}(X_k^m + C_k^m)$.

- **Quasi-ML Nonlinear Detection**: For this method the PAR reduction signal is simply the nonlinear characteristic of the transmitter nonlinearity, which the receiver must estimate.

Chapter 4

PAR REDUCTION BY TONE RESERVATION

T
HIS CHAPTER DESCRIBES THE FIRST NEW FAMILY of methods to reduce the PAR of multicarrier signals. Most of the ideas presented here were first described in a number of standards contributions [Tellado and Cioffi, 1997b, Tellado and Cioffi, 1998d, Tellado and Cioffi, 1998g, Tellado and Cioffi, 1998h] and in a patent application [Tellado and Cioffi, 1997a], and more recently in [Tellado and Cioffi, 1998a, Tellado and Cioffi, 1998f]. All these methods are based on the general additive model of PAR reduction described in Section 9. of Chapter 3, and more specifically on the Tone Reservation concept. The basic idea is to *reserve* a small set of tones for PAR reduction. Fortunately, the problem of computing the values for these reserved tones that minimize the PAR can be formulated as a convex problem and can be solved exactly. We also show that suboptimal gradient techniques converge fast to this optimal solution. The amount of PAR reduction depends on the number of reserved tones, their locations within the frequency vector and the amount of complexity.

Section 3. shows that the resultant formulation can be minimized by convex optimization techniques that lead to large reduction in PAR. Section 5. describes some efficient methods that achieve 4-5 dB reduction in PAR with minimal rate loss and require only $\mathcal{O}(N)$ operations per DMT/OFDM symbol. We would like to note some excellent simultaneous work by A. Gatherer and M. Polley. In [Gatherer and Polley, 1997], they independently derived a Projection algorithm that is a special case of the gradient methods in Section 4. The methods presented here, however, show significant additional improvement to that particular method, most notably by optimizing the peak reduction subspace, and by a more careful design of the peak reduction kernel.

1.   PROBLEM FORMULATION

Section 4. of Chapter 3 showed that the oversampled representations of continuous-time multicarrier symbols were accurate predictors of continuous-time PAR. Thus, in all the mathematical derivations in this chapter, only the discrete-time symbols will be considered. The IDFT modulator operation of (3.6) is:

$$x^m[n/L] \quad = \quad \frac{1}{\sqrt{N}} \sum_{k=-\frac{N}{2}}^{\frac{N}{2}-1} X_k^m e^{j2\pi kn/NL} w[n/L] \tag{4.1}$$

$$= \quad IDFT(\sqrt{L}\, \mathbf{X}_L^m) \tag{4.2}$$

which can also be written in matrix notation as

$$\begin{bmatrix} x_{0/L}^m \\ x_{1/L}^m \\ x_{2/L}^m \\ \vdots \\ \vdots \\ x_{n/L}^m \\ \vdots \\ \vdots \\ x_{(NL-1)/L}^m \end{bmatrix}_{NL \times 1} = \quad Q \begin{bmatrix} X_0^m \\ \vdots \\ X_{N/2-1}^m \\ 0 \\ \vdots \\ 0 \\ X_{N/2}^m \\ \vdots \\ X_{N-1}^m \end{bmatrix}_{NL \times 1} \tag{4.3}$$

where Q is the IDFT matrix of size NL but scaled by $\sqrt{L}$. That is,

$$Q = \frac{1}{\sqrt{N}} \begin{bmatrix} 1 & 1 & \cdots & 1 \\ 1 & e^{j\frac{2\pi}{NL}1\cdot1} & \cdots & e^{j\frac{2\pi}{NL}1(NL-1)} \\ 1 & e^{j\frac{2\pi}{NL}2\cdot1} & \cdots & e^{j\frac{2\pi}{NL}2(NL-1)} \\ \vdots & \vdots & \ddots & \vdots \\ 1 & e^{j\frac{2\pi}{NL}n\cdot1} & \cdots & e^{j\frac{2\pi}{NL}n(NL-1)} \\ \vdots & \vdots & \ddots & \vdots \\ 1 & e^{j\frac{2\pi}{NL}(NL-1)1} & \cdots & e^{j\frac{2\pi}{NL}(NL-1)(NL-1)} \end{bmatrix}_{NL \times NL} \tag{4.4}$$

Using this notation the oversampled multicarrier modulator can be expressed compactly as $\mathbf{x}_L^m = Q\mathbf{X}_L^m$, where the elements of Q are $q_{n,k} = \frac{1}{\sqrt{N}} e^{j2\pi nk/NL}$. Since the middle $N(L-1)$ terms of $\mathbf{X}_L^m$ are always zero, this matrix is redundant. Calling Q_L the submatrix of Q formed by selecting its first and last $N/2$ columns, the oversampled IDFT can be expressed as

$$
\mathbf{x}_L^m =
\begin{bmatrix}
x_{0/L}^m \\
x_{1/L}^m \\
\vdots \\
x_{n/L}^m \\
\vdots \\
x_{(NL-1)/L}^m
\end{bmatrix}_{NL\times 1}
= Q_L
\begin{bmatrix}
X_0^m \\
X_1 \\
\vdots \\
X_k^m \\
\vdots \\
X_{N-1}^m
\end{bmatrix}_{N\times 1}
= Q_L \mathbf{X}_L^m \qquad (4.5)
$$

This new matrix Q_L has dimensions $NL \times N$. For $L = 1$, Q_L is the standard IDFT matrix of size N. In the following derivations, the notation $\mathbf{q}_{n,L}^{row}$ denotes the n-th row of Q_L i.e.

$$
\mathbf{q}_{n,L}^{row} = [1 \quad e^{j\frac{2\pi}{NL}n\cdot 1} \quad \cdots \quad e^{j\frac{2\pi}{NL}n(\frac{N}{2}-1)} \quad e^{j\frac{2\pi}{NL}n(NL-\frac{N}{2})} \quad \cdots \quad e^{j\frac{2\pi}{NL}n(NL-1)}]_{N\times 1}
$$
$$(4.6)$$

and $\mathbf{q}_{k,L}^{col}$ denotes the k-th column of Q_L. That is,

$$
\mathbf{q}_{k,L}^{col} =
\begin{bmatrix}
1 \\
e^{j\frac{2\pi}{NL}1\cdot k} \\
e^{j\frac{2\pi}{NL}2\cdot k} \\
\vdots \\
e^{j\frac{2\pi}{NL}(NL-1)\cdot k}
\end{bmatrix}_{NL\times 1}
\qquad (4.7)
$$

for $0 \le k < N/2$ and

$$
\mathbf{q}_{k,L}^{col} =
\begin{bmatrix}
1 \\
e^{j\frac{2\pi}{NL}1(N(L-1)+k)} \\
e^{j\frac{2\pi}{NL}2(N(L-1)+k)} \\
\vdots \\
e^{j\frac{2\pi}{NL}(NL-1)(N(L-1)+k)}
\end{bmatrix}_{NL\times 1}
=
\begin{bmatrix}
1 \\
e^{-j\frac{2\pi}{NL}1\cdot(N-k)} \\
e^{-j\frac{2\pi}{NL}2\cdot(N-k)} \\
\vdots \\
e^{-j\frac{2\pi}{NL}(NL-1)\cdot(N-k)}
\end{bmatrix}_{NL\times 1}
\qquad (4.8)
$$

for $N/2 \leq k < N$. It is easy to show that $\mathbf{q}^{col}_{N-k,L} = (\mathbf{q}^{col}_{k,L})^*$. In a similar manner, the fundamental additive PAR reduction equation (3.135) can be expressed in matrix notation as:

$$
\begin{bmatrix} x^m_{0/L} \\ x^m_{1/L} \\ x^m_{2/L} \\ \vdots \\ x^m_{n/L} \\ \vdots \\ x^m_{\frac{NL-1}{L}} \end{bmatrix}_{NL \times 1} + \begin{bmatrix} c^m_{0/L} \\ c^m_{1/L} \\ c^m_{2/L} \\ \vdots \\ c^m_{n/L} \\ \vdots \\ c^m_{\frac{NL-1}{L}} \end{bmatrix}_{NL \times 1} = Q_L \left(\begin{bmatrix} X^m_0 \\ X^m_1 \\ \vdots \\ X^m_k \\ \vdots \\ X^m_{N-1} \end{bmatrix} + \begin{bmatrix} C^m_0 \\ C^m_1 \\ \vdots \\ C^m_k \\ \vdots \\ C^m_{N-1} \end{bmatrix} \right)_{N \times 1}
$$

or more compactly as

$$
\bar{\mathbf{x}}^m_L = \mathbf{x}^m_L + \mathbf{c}^m_L = Q_L(\mathbf{X}^m + \mathbf{C}^m) \tag{4.9}
$$

For baseband transmission, both $\mathbf{x}^m_L$ and $\mathbf{c}^m_L$ must be real valued sequence. Therefore, $\mathbf{X}^m$ and $\mathbf{C}^m$ must possess the Hermitian symmetry properties. For example, if N is even, X^m_k and C^m_k must satisfy $X^m_k = (X^m_{N-k})^*$ and $C_k = (C^m_{N-k})^*$. Moreover X^m_0, $X^m_{N/2}$, C^m_0 and $C^m_{N/2}$ must be real valued. Similar constraints are necessary when N is odd.

2.   PAR REDUCTION SIGNALS FOR TONE RESERVATION

The receiver must decode the values in $\mathbf{X}^m$ from the received vector $\mathbf{X}^m + \mathbf{C}^m$. Certain structures in the design of $\mathbf{C}^m$ can lead to very simple transmitters and receivers. This chapter describes the first novel PAR reduction method called Tone Reservation.

In the Tone Reservation method, both the transmitter and the receiver agree on reserving a small subset of tones for generating PAR reduction signals. These reserved tones are not used for data transmission. By constraining the PAR reduction signals to lie in the subspace generated by these reserved tones, the data vector $\mathbf{X}^m$ and the PAR reduction vector $\mathbf{C}^m$ can be easily separated since they are nonzero in disjoint or mutually exclusive frequency bins, i.e. $X^m_k C^m_k = 0$. Let $\mathcal{R} = \{i_0, \ldots, i_{R-1}\}$ denote the ordered subset of tones that are reserved for PAR reduction. The indices satisfy $0 \leq i_0 < \ldots < i_r < \ldots < i_{R-1} < N$ where $R << N$ is the number of reserved tones. Calling $\mathcal{N} = \{0, 1, \ldots, N-1\}$ the set of all tones in the multicarrier symbol and $\mathcal{R}^c$ the complement of $\mathcal{R}$ in $\mathcal{N}$, i.e. $\mathcal{N} = \mathcal{R} + \mathcal{R}^c$, the tone values satisfy

$$X_k^m + C_k^m = \begin{cases} C_k^m, & k \in \mathcal{R} \\ X_k^m, & k \in \mathcal{R}^c \end{cases} \qquad (4.10)$$

The multicarrier modulator in (3.135) can be rewritten as

$$x^m[n/L] + c^m[n/L] = \frac{1}{\sqrt{N}} \sum_{k \in \mathcal{R}^c} X_k^m e^{j2\pi kn/NL} + \frac{1}{\sqrt{N}} \sum_{k \in \mathcal{R}} C_k^m e^{j2\pi kn/NL} \qquad (4.11)$$

The set of reserved tones in $\mathcal{R}$, where the PAR reduction signals are nonzero, will also be called Peak Reduction Tones (PRT).

With this choice of $\mathbf{X}^m$ and $\mathbf{C}^m$, the demodulator is very simple. Assuming the cyclic prefix is longer than the channel impulse response, from (2.20), the output of the receiver DFT is

$$H_k(X_k^m + C_k^m) + N_k^m = \begin{cases} H_k C_k^m + N_k^m, & k \in \mathcal{R} \\ H_k X_k^m + N_k^m, & k \in \mathcal{R}^c \end{cases} \qquad (4.12)$$

where H_k is the DFT of the channel impulse response and N_k^m is the DFT of the additive noise. The only change at the receiver is that it must only decode the values in tones $k \in \mathcal{R}^c$ and ignore the PRT values[1]. This additive PAR reduction structure for dividing the tones is described graphically in Figure 4.1.

Let's call $\hat{\mathbf{C}}^m$ the vector of length R obtained by selecting the values with index $k \in \mathcal{R}$ from $\mathbf{C}^m$, i.e. $\hat{\mathbf{C}}^m = [C_{i_0}^m \cdots C_{i_{R-1}}^m]^T$, and similarly $\hat{Q}_L$ the submatrix of Q_L constructed by choosing its columns $i_r \in \mathcal{R}$, i.e. $\hat{Q}_L = [\mathbf{q}_{i_0,L}^{col} | \cdots | \mathbf{q}_{i_{R-1},L}^{col}]$. Then,

$$\mathbf{c}^m = Q_L \mathbf{C}^m = \hat{Q}_L \hat{\mathbf{C}}^m \qquad (4.13)$$

since the remaining values of $\mathbf{C}^m$ are zero. Using (4.13), the general additive PAR reduction expression in (4.9) can be simplified to

$$\bar{\mathbf{x}}_L^m = \mathbf{x}_L^m + \mathbf{c}_L^m = \mathbf{x}_L^m + \hat{Q}_L \hat{\mathbf{C}}^m \qquad (4.14)$$

---

[1] The values in the reserved tones are a function of $\mathbf{X}^m$ and therefore they also contain some information about the data, but this information will be ignored for simplicity.

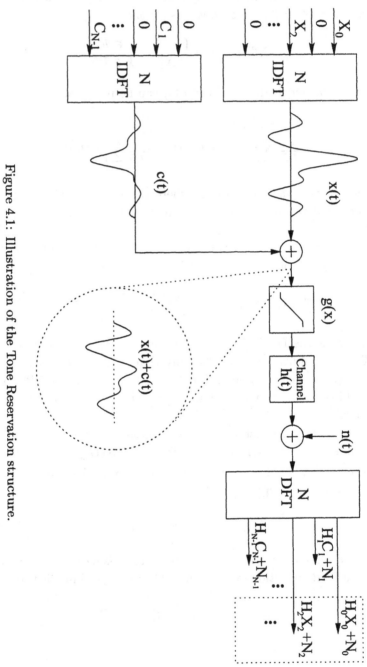

Figure 4.1: Illustration of the Tone Reservation structure.

Restricting R values of X_k to be zero, implies no information can be transmitted in these tones. Therefore, the *Tone Reservation Rate (TRR)* is R/N. Denoting b_k as the number of bits transmitted in tone k, the number of bits lost for Tone Reservation PAR reduction is $\sum_{r=0}^{R-1} b_{i_r}$. Thus, the *Data Rate Loss (DRL)* is

$$DRL = \frac{\sum_{r=0}^{R-1} b_{i_r}}{\sum_{k=0}^{N-1} b_k} \qquad (4.15)$$

In many applications such as ADSL or VDSL, the N tones in our multicarrier symbol do not have equal SNR_k and the values of b_k can differ greatly from tone to tone. By choosing the set $\{i_0, \ldots, i_{R-1}\}$ such that the values of b_{i_r} are small, we can get $DRL \ll TRR$. In particular, if $b_{i_r} = 0$, then there is no data rate loss on that tone if it is reserved for PAR reduction. For applications where $b_k = constant$, e.g. OFDM, R needs to be minimized to reduce the DRL.

3.   OPTIMAL PAR REDUCTION SIGNALS FOR TONE RESERVATION

Using the compact vector notation, the PAR of the *m-th* symbol $\mathbf{x}_L^m$ prior to PAR reduction is defined as:

$$PAR\{\mathbf{x}_L^m\} = \frac{\|\mathbf{x}_L^m\|_\infty^2}{E\{|x^m[n/L]|^2\}} \qquad (4.16)$$

where $\|\mathbf{v}\|_\infty$ denotes the ∞-*norm*[2] of the vector $\mathbf{v}$. Following (3.136), the PAR of the additive symbol $\mathbf{x}_L^m + \mathbf{c}_L^m$ is defined as:

$$PAR\{\mathbf{x}_L^m + \mathbf{c}_L^m\} = \frac{\|\mathbf{x}_L^m + \mathbf{c}_L^m\|_\infty^2}{E\{|x^m[n/L]|^2\}} \qquad (4.17)$$

Since the denominator is not a function of the PAR reduction signals, the problem of minimizing the PAR of the combined signal is equivalent to computing the value of $\mathbf{c}^{m,opt}$, or equivalently $\mathbf{C}^{m,opt}$, that minimizes

---

[2]The ∞-*norm* of a vector is the maximum of the absolute values of its components

the maximum peak value or ∞-*norm* of $\mathbf{x}_L^m + \mathbf{c}_L^m$. That is,

$$\min_{\mathbf{c}_L} \|\mathbf{x}_L^m + \mathbf{c}_L\|_\infty = \min_{\hat{\mathbf{C}}} \|\mathbf{x}_L^m + \hat{Q}_L \hat{\mathbf{C}}\|_\infty \qquad (4.18)$$

This optimization problem is convex in the vector of variables $\hat{\mathbf{C}} = [C_{i_0} \cdots C_{i_r} \cdots C_{i_{R-1}}]^T$. First, (4.18) is equivalent to:

$$\min_{\hat{\mathbf{C}}} \quad t$$

$$\text{subject to:} \quad \|\mathbf{x}_L^m + \hat{Q}_L \hat{\mathbf{C}}\|_\infty^2 \leq t \qquad (4.19)$$

Denoting $\hat{\mathbf{q}}_{n,L}^{row}$ the *n-th* row of $\hat{Q}_L$, the optimization above is equivalent to:

$$\min_{\hat{\mathbf{C}}} \quad t$$

$$\text{subject to:} \quad |x_{0/L}^m + \hat{\mathbf{q}}_{0,L}^{row} \hat{\mathbf{C}}|^2 \leq t$$

$$|x_{1/L}^m + \hat{\mathbf{q}}_{1,L}^{row} \hat{\mathbf{C}}|^2 \leq t$$

$$\vdots \qquad\qquad (4.20)$$

$$|x_{n/L}^m + \hat{\mathbf{q}}_{n,L}^{row} \hat{\mathbf{C}}|^2 \leq t$$

$$\vdots$$

$$|x_{(NL-1)/L}^m + \hat{\mathbf{q}}_{NL-1,L}^{row} \hat{\mathbf{C}}|^2 \leq t$$

By expanding the square of the absolute values, the optimization problem becomes,

$$\min_{\hat{\mathbf{C}}} \quad t$$

$$\text{subject to:} \quad (x_{\frac{0}{L}}^m + \hat{\mathbf{q}}_{0,L}^{row} \hat{\mathbf{C}})^* (x_{\frac{0}{L}}^m + \hat{\mathbf{q}}_{0,L}^{row} \hat{\mathbf{C}}) \leq t$$

$$\vdots$$

$$(x_{\frac{n}{L}}^m + \hat{\mathbf{q}}_{n,L}^{row} \hat{\mathbf{C}})^* (x_{\frac{n}{L}}^m + \hat{\mathbf{q}}_{n,L}^{row} \hat{\mathbf{C}}) \leq t \qquad (4.21)$$

$$\vdots$$

$$(x_{\frac{NL-1}{L}}^m + \hat{\mathbf{q}}_{NL-1,L}^{row} \hat{\mathbf{C}})^* (x_{\frac{NL-1}{L}}^m + \hat{\mathbf{q}}_{NL-1,L}^{row} \hat{\mathbf{C}}) \leq t$$

This is a convex problem since it minimizes a linear constraint over an intersection of quadratic (and thus convex) constraints on the variables $\hat{\mathbf{C}}$. Specifically, this is a special case of a Quadratically Constrained

Quadratic Program (QCQP) which is a well studied convex problem [Boyd and Vandenberghe, 1997].

For baseband multicarrier transmission, the QCQP reduces to a Linear Program (LP). First of all, real multicarrier signals require real PAR reduction signals. Since $\mathbf{x}_L^m + \mathbf{c}_L^m$ must be real valued, $\mathbf{C}^m$ must satisfy the Hermitian symmetry property. The following derivation will focus on the case when N is even but can be easily repeated if N is odd. From Section 1., if N is even, C_k^m must satisfy $C_k = (C_{N-k}^m)^*$ and C_0^m and $C_{N/2}^m$ must be real valued. If $i_r \in \mathcal{R}$, the Hermitian symmetry property requires[3] $N - i_r \in \mathcal{R}$. To simplify the derivation, the special cases where $i_r = 0$ and $i_r = N/2$ will be excluded, i.e. $0 \notin \mathcal{R}$ and $N/2 \notin \mathcal{R}$. For this case, the PRT index set satisfies $i_r = N - i_{R-r}$ and $\mathcal{R}$ has the following structure:

$$\mathcal{R} = \{i_0, \ldots, i_r, \ldots, i_{R/2-1}, N - i_{R/2-1}, \ldots, N - i_r, \ldots, N - i_0\} \tag{4.22}$$

The first $R/2$ elements in $\mathcal{R}$, which are denoted by

$$\mathcal{R}_{1/2} = \{i_0, \ldots, i_r, \ldots, i_{R/2-1}\} \tag{4.23}$$

completely determine the PRT set. Thus, the PAR reduction term in (4.11) reduces to

$$c^m[n/L] = \frac{1}{\sqrt{N}} \sum_{r=0}^{R-1} C_{i_r}^m e^{j2\pi i_r n/NL} = \tag{4.24}$$

$$= \frac{1}{\sqrt{N}} \left(\sum_{r=0}^{R/2-1} C_{i_r}^m e^{j2\pi i_r n/NL} + \sum_{r=R/2}^{R-1} C_{i_r}^m e^{j2\pi i_r n/NL} \right) \tag{4.25}$$

$$= \frac{1}{\sqrt{N}} \left(\sum_{r=0}^{R/2-1} C_{i_r}^m e^{j2\pi i_r n/NL} + \sum_{r=0}^{R/2-1} C_{N-i_r}^m e^{-j2\pi i_r n/NL} \right) \tag{4.26}$$

$$= \frac{1}{\sqrt{N}} \sum_{r=0}^{R/2-1} \left(C_{i_r}^m e^{j2\pi i_r n/NL} + (C_{i_r}^m e^{j2\pi i_r n/NL})^* \right) \tag{4.27}$$

$$= \frac{2}{\sqrt{N}} \sum_{r=0}^{R/2-1} \left(C_{i_r,re}^m \cos(2\pi i_r n/NL) - C_{i_r,im}^m \sin(2\pi i_r n/NL) \right) \tag{4.28}$$

---

[3]The indices $i_r = 0$ and $i_r = N/2$ coincide with their symmetric index.

where $n = 0, \ldots, NL - 1$, and $C_{i_r,re}^m$ and $C_{i_r,im}^m$ are the real and imaginary parts of $C_{i_r}^m$. These NL equations can be also expressed in matrix notation

$$
\begin{bmatrix}
c_{0/L}^m \\
c_{1/L}^m \\
c_{2/L}^m \\
\vdots \\
c_{n/L}^m \\
\vdots \\
c_{(NL-1)/L}^m
\end{bmatrix}_{NL \times 1}
= \check{Q}_L
\begin{bmatrix}
C_{i_0,re}^m \\
C_{i_0,im}^m \\
\vdots \\
C_{i_r,re}^m \\
C_{i_r,im}^m \\
\vdots \\
C_{i_{R/2-1},re}^m \\
C_{i_{R/2-1},im}^m
\end{bmatrix}_{R \times 1}
\tag{4.29}
$$

All elements in (4.29) are real valued and $\check{Q}_L$ denotes the $NL \times R$ matrix that includes all the sinusoidal terms in (4.28). Calling $\check{C}^m$ the vector on the right-hand side, (4.29) can be expressed compactly as $c^m = \check{Q}_L \check{C}^m$. With this notation, minimizing the PAR in (4.17) for a baseband multicarrier system is equivalent to solving:

$$
\min_{c_L} \| x_L^m + c_L \|_\infty = \min_{\check{C}} \| x_L^m + \check{Q}_L \check{C} \|_\infty
\tag{4.30}
$$

Equation (4.30) can be recast as a LP and can be solved exactly. First, (4.30) is equivalent to:

$$
\min_{\check{C}} \quad t
$$
$$
\text{subject to}: \quad |x_{n/L} + \check{q}_{n,L}^{row} \check{C}| \leq t, \quad n = 0, \ldots, NL - 1. \tag{4.31}
$$

where $\check{q}_{n,L}^{row}$ is the n-th row of $\check{Q}_L$. These NL scalar constraints can be written in vector form, as follows:

$$
\min_{\check{C}} \quad t
$$
$$
\text{subject to}: \quad x_L^m + \check{Q}_L \check{C} \leq t1_{NL},
\tag{4.32}
$$
$$
x_L^m + \check{Q}_L \check{C} \geq -t1_{NL},
$$

where $\mathbf{1}_{NL}$ is a column vector with NL ones and $\mathbf{y} \leq \mathbf{z}$ indicates a component-wise vector inequality, i.e. $y_n \leq z_n$, $\forall n$. Moving the unknowns $\check{\mathbf{C}}$ and t to the left hand side,

$$\min_{\check{\mathbf{C}}} \quad t$$
$$\text{subject to}: \quad \check{Q}_L \check{\mathbf{C}} - t\mathbf{1}_{NL} \leq -\mathbf{x}_L^m, \qquad (4.33)$$
$$\check{Q}_L \check{\mathbf{C}} + t\mathbf{1}_{NL} \geq -\mathbf{x}_L^m,$$

and grouping the constraints, the optimization problem can be reformulated as

$$\min_{\check{\mathbf{C}}} \quad t$$
$$\text{subject to}: \quad \begin{pmatrix} \check{Q}_L & -\mathbf{1}_{NL} \\ -\check{Q}_L & -\mathbf{1}_{NL} \end{pmatrix} \begin{pmatrix} \check{\mathbf{C}} \\ t \end{pmatrix} \leq \begin{pmatrix} -\mathbf{x}_L^m \\ \mathbf{x}_L^m \end{pmatrix} \quad (4.34)$$

This LP has $R+1$ unknowns $\{\check{\mathbf{C}}, t\}$ and $2NL$ inequalities and is expressed in the standard LP form [Boyd and Vandenberghe, 1997]:

$$\min_{\mathbf{y}} \quad \mathbf{c}^T \mathbf{y}$$
$$\text{subject to}: \quad A\mathbf{y} \leq \mathbf{b} \qquad (4.35)$$

where $\mathbf{y}$ are the optimization variables. The matrix A, and the vectors $\mathbf{b}$ and $\mathbf{c}$ are known parameters.

Convex optimization problems and in particular Linear Programs have been studied extensively. In addition to having a unique solution, their convergence properties are well known and many algorithms provide lower bounds on the solution at every iteration [Boyd and Vandenberghe, 1997]. Both LP and QCQP are two of the simplest and better documented convex problems. For a general LP of this size, the complexity is $\mathcal{O}(RLN^2)$. In this case, the LP is very structured as $\check{Q}_L$ is related to the IDFT matrix. Therefore, using this structure this LP can be solved with complexity $\mathcal{O}(N \log N)$ [Boyd et al., 1994]. These complexity counts are necessary only if the exact solution to (4.18) is desired using general purpose algorithms. Section 5., describes iterative gradient algorithms that give good approximations to (4.18) with complexity $\mathcal{O}(N)$.

Figure 4.2 shows the distribution of the PAR for both the original multicarrier symbols $\mathbf{x}_L^m$, and the peak-optimized symbols $\mathbf{x}_L^m + \mathbf{c}_L^{m,opt}$, where $\mathbf{c}_L^{m,opt}$ is obtained by solving (4.30) exactly. All PAR CCDF plots can be used to measure the clipping probability, i.e. the probability that the symbols exceed a given PAR threshold, PAR_0. For example, in Figure 4.2, if the symbol CCDF PAR follows the solid line, and the transmitter can only tolerate $PAR_0 = 12\ dB$, then 3% of the transmitted symbols will be clipped, i.e. $Prob\{PAR_m > 12\ dB\} = .03$. The solid line in Figure 4.2 shows the clipping probability curve when no clipping reduction is performed. The dashed-dotted line in Figure 4.2 plots the clipping probability after peak-reduction with $R/N = 5\%$. For this case, if the desired clipping probability is less than 10^{-5}, the PAR can be reduced from 15 dB to 9 dB. In other words, the statistical PAR has been reduced by 6 dB if we can only tolerate a clipping probability below 10^{-5}. The dashed line in Figure 4.2 plots results for $R/N = 20\%$. In this case, PAR decreases from 15 dB to 5 dB for a clipping probability below 10^{-5}.

Adding $\mathbf{c}_L^{m,opt}$ to $\mathbf{x}_L^m$ reduces the PAR but also increases the transmit power slightly. The relative mean power increase ΔE, due to PAR reduction is defined as:

$$\Delta E = 10 \log_{10} \frac{E\{\|\mathbf{x}_L^m + \mathbf{c}_L^m\|_2^2\}}{E\{\|\mathbf{x}_L^m\|_2^2\}} \tag{4.36}$$

For the Tone Reservation case, $\mathbf{x}_L^m$ and $\mathbf{c}_L^m$ are orthogonal. Thus,

$$E\{\|\mathbf{x}_L^m + \mathbf{c}_L^m\|_2^2\} = E\{\|\mathbf{x}_L^m\|_2^2\} + E\{\|\mathbf{c}_L^m\|_2^2\} \tag{4.37}$$

and ΔE simplifies to

$$\Delta E = 10 \log_{10} \left(1 + \frac{E\{\|\mathbf{c}_L^m\|_2^2\}}{E\{\|\mathbf{x}_L^m\|_2^2\}} \right) \tag{4.38}$$

For the example in Figure 4.2, the optimal Tone Reservation solution, $\mathbf{c}_L^{m,opt}$, increases the mean power by $\Delta E = 1\ dB$ for $R/N = 5\%$, and $\Delta E = 0.5\ dB$ for $R/N = 20\%$.

These results show that the Tone Reservation method can significantly reduce the PAR by solving (4.18) or (4.30) *exactly* with small increases of the average transmit power. The following sections show that most of the PAR reduction can be achieved from low complexity iterative methods

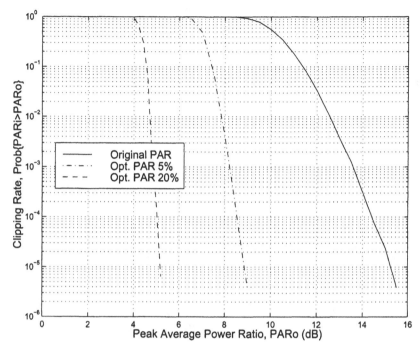

Figure 4.2: CCDF of $PAR\{x[n] + c[n]\}$ for $N = 512$ when $\frac{R}{N} = 5\%$ and $\frac{R}{N} = 20\%$ with Randomly-Optimized set $\mathcal{R}^{\star}$.

after a small number of iterations. Moreover, the power increase is much smaller for these simplified solutions.

4. SIMPLE GRADIENT ALGORITHMS WITH FAST CONVERGENCE

The previous sections formulated the Tone Reservation additive PAR reduction method and described the exact solution to the PAR minimization problem. The optimal solution is obtained by solving a QCQP for the complex multicarrier case (passband applications), or by solving a LP for the real multicarrier case (baseband transmission). This section describes a simple iterative algorithm for computing the Tone Reservation PAR reduction signal based on taking the gradient of the Mean Square Error (MSE) of the transmitter distortion function.

Section 5. of Chapter 3 introduced the concept of memoryless non-linearity, and described a number of commonly accepted models. The notation $g(\cdot)$ was used to represent a general nonlinearity. Moreover, the distortion signal for the *m-th* symbol was defined as $d^{(\mathbf{X},g)} = g(\mathbf{x}_L^m) - \mathbf{x}_L^m$. Therefore, the total distortion over the nonperiodic part of a multicarrier symbol is:

$$\|\mathbf{x}_L^m - g(\mathbf{x}_L^m)\|_2^2 = \sum_{n=0}^{NL-1} (x[n/L] - g(x[n/L]))^2 \qquad (4.39)$$

and the Signal to Distortion power Ratio (SDR) is:

$$SDR = \frac{\|\mathbf{x}_L^m\|_2^2}{\|\mathbf{x}_L^m - g(\mathbf{x}_L^m)\|_2^2} \qquad (4.40)$$

By adding the PAR reduction signal to the transmit symbol, the transmitted time sequence is $\mathbf{x}_L^m + \hat{Q}_L\hat{\mathbf{C}}$ and the SDR will be modified as follows:

$$SDR = \frac{\|\mathbf{x}_L^m\|_2^2}{\|\mathbf{x}_L^m + \hat{Q}_L\hat{\mathbf{C}} - g(\mathbf{x}_L^m + \hat{Q}_L\hat{\mathbf{C}})\|_2^2} \qquad (4.41)$$

In the previous sections, the target was to minimize the PAR of $\mathbf{x}_L^m + \hat{Q}_L\hat{\mathbf{C}}$. For a general nonlinearity, a more natural approach is to maximize the SDR, which is a more accepted measure of the quality of our transmit signal. Since the numerator of (4.41) is a constant, this optimization problem is equivalent to minimizing the denominator

$$\min_{\hat{\mathbf{C}}} \|\mathbf{x}_L^m + \hat{Q}_L\hat{\mathbf{C}} - g(\mathbf{x}_L^m + \hat{Q}_L\hat{\mathbf{C}})\|_2^2 \qquad (4.42)$$

If the nonlinear function $x_\tau - g(x_\tau)$ is a convex function (as is the case for the SL and SSPA nonlinearities), then the minimization problem of (4.42) is also convex since it is a composition of convex functions in the unknowns $\hat{\mathbf{C}}$. Thus, the values of $\hat{\mathbf{C}}$ that maximize the SDR can be computed exactly. The rest of this section will focus on the important special case of a SL nonlinearity with maximum amplitude A, since most nonlinear behaviors can be reduced to the SL case with standard predistortion techniques. Moreover it also simplifies the discussion. For the SL nonlinearity case, the more commonly used term of Signal to Clipping noise power Ratio (SCR) will be used to denote SDR.

In all the following, instead of solving for $\hat{\mathbf{C}}$ exactly, we will describe a gradient based iterative algorithm with complexity $\mathcal{O}(N)$.

It can be shown that the gradient with respect to $\hat{\mathbf{C}}$ of (4.42) is:

$$\nabla_{\hat{\mathbf{C}}} \|\mathbf{x}_L^m + \hat{Q}_L\hat{\mathbf{C}} - g_{sL}(\mathbf{x}_L^m + \hat{Q}_L\hat{\mathbf{C}})\|_2^2 =$$
$$2 \sum_{|x_n^m + c_n| > A} (x_n^m + c_n - Ae^{j\arg\{x_n^m + c_n\}})(\hat{\mathbf{q}}_{n,L}^{row})^* \quad (4.43)$$

where the oversampling factor L has been dropped from the sample subindeces $x_{n/L}^m$ and $c_{n/L}$ to simplify the notation. Starting with the initial conditions of $\hat{\mathbf{C}}^{m(0)} = \mathbf{0}_R$ and $\mathbf{c}^{m(0)} = \mathbf{0}_{NL}$, where $\mathbf{0}_M$ represents the all zero column vector of length M, the following gradient iterative algorithm can be constructed to update the PAR reduction vector $\hat{\mathbf{C}}$:

$$\hat{\mathbf{C}}^{m(i+1)} = \hat{\mathbf{C}}^{m(i)} - \frac{\mu}{2}\nabla_{\hat{\mathbf{C}}}\|\mathbf{x}_L^m + \hat{Q}_L\hat{\mathbf{C}}^{m(i)} - g_{sL}(\mathbf{x}_L^m + \hat{Q}_L\hat{\mathbf{C}}^{m(i)})\|_2^2 \quad (4.44)$$

or equivalently:

$$\hat{\mathbf{C}}^{m(i+1)} = \hat{\mathbf{C}}^{m(i)} - \mu \sum_{|x_n^m + c_n^{m(i)}| > A} (x_n^m + c_n^{m(i)} - Ae^{j\arg\{x_n^m + c_n^{m(i)}\}})(\hat{\mathbf{q}}_{n,L}^{row})^* \quad (4.45)$$

In this equation, both the time domain $\mathbf{c}^{m(i)}$ and frequency domain $\hat{\mathbf{C}}^{m(i+1)}$ PAR reduction signals must be computed. The algorithm can be simplified by updating the PAR reduction signal in the time domain only. This is achieved by pre-multiplying both sides of (4.45) by $\hat{Q}_L$, which is the IDFT matrix for vectors with nonzero elements in $\mathcal{R}$.

$$\mathbf{c}^{m(i+1)} = \mathbf{c}^{m(i)} - \mu \sum_{|x_n^m + c_n^{m(i)}| > A} (x_n^m + c_n^{m(i)} - Ae^{j\arg\{x_n^m + c_n^{m(i)}\}})\hat{Q}_L(\hat{\mathbf{q}}_{n,L}^{row})^*$$
$$(4.46)$$

This new time domain update never requires the computation of $\hat{\mathbf{C}}^{m(i+1)}$. Moreover, unlike most low-overhead PAR reduction methods in the literature, no DFTs are necessary to compute $\mathbf{c}^{m(i)}$. With the gradient update in (4.46), the PAR reduction algorithm is

$$\bar{\mathbf{x}}_L^{m(i+1)} = \bar{\mathbf{x}}_L^{m(i)} - \mu \sum_{|\bar{x}_n^{m(i)}| > A} \underbrace{(\bar{x}_n^{m(i)} - Ae^{j\arg\{\bar{x}_n^{m(i)}\}})}_{\alpha_n^{m(i)}} \underbrace{\hat{Q}_L(\hat{\mathbf{q}}_{n,L}^{row})^*}_{\mathbf{p}_n^{\|2}} \quad (4.47)$$

where $\bar{\mathbf{x}}_L^{m(i)} = \mathbf{x}_L^m + \mathbf{c}_L^{m(i)}$.

The apparent complexity of (4.47) is misleading, since the term $\alpha_n^{m(i)} = \bar{x}_n^{m(i)} - Ae^{j\arg\{\bar{x}_n^{m(i)}\}}$ is just a complex scalar, and the term $\mathbf{p}_n^{\|\|^2} = \hat{Q}_L(\hat{\mathbf{q}}_{n,L}^{row})^*$ is a constant vector that only depends on $\mathcal{R}$ and thus can be precomputed and stored in memory during the modem initialization. The full Tone Reservation iterative algorithm based on the SCR gradient is simply

$$\bar{\mathbf{x}}_L^{m(i+1)} = \bar{\mathbf{x}}_L^{m(i)} - \mu \sum_{|\bar{x}_n^{m(i)}| > A} \alpha_n^{m(i)} \mathbf{p}_n^{\|\|^2} \tag{4.48}$$

where $\bar{\mathbf{x}}_L^{m(0)} = \mathbf{x}_L^m$. This equation also applies to the real multicarrier case with the following straightforward modifications:

$$\alpha_n^{m(i)} = \bar{x}_n^{m(i)} - A \cdot sign\{\bar{x}_n^{m(i)}\} \tag{4.49}$$

$$\mathbf{p}_n^{\|\|^2} = \check{Q}_L(\check{\mathbf{q}}_{n,L}^{row})^* \tag{4.50}$$

Since the condition $|\bar{x}_{n_i}^{m(i)}| > A$ can occur for any of the NL samples of the multicarrier symbol, the transmitter must compute and store all possible kernels $\mathbf{p}_n^{\|\|^2}$, for $n = 0, \dots, NL - 1$. Fortunately, the kernels satisfy the circular shift property:

$$p_{n_i}^{\|\|^2}[n/L] = p_0^{\|\|^2}[((n - n_i))_{NL}/L] \tag{4.51}$$

where $p[((n - n_i))_{NL}/L]$ represents the circular shifted sequence:

$$[p[n_i/L] \ p[(n_i + 1)/L] \ \dots \ p[(NL - 1)/L] \ p[0/L] \ \dots \ p[(n_i - 1)/L]]^T$$

This follows from the circular shift property of the DFT, since $\hat{Q}_L$ is the IDFT matrix over the subspace of tones $\mathcal{R}$. It is easy to show that

$$\hat{\mathbf{q}}_{n_i,L}^{row} = \hat{\mathbf{q}}_{n_i-1,L}^{row} \odot [e^{j\frac{2\pi}{NL}i_0} \ \dots \ e^{j\frac{2\pi}{NL}i_r} \ \dots \ e^{j\frac{2\pi}{NL}i_{R-1}}]_{1 \times R} \tag{4.52}$$

where $\odot$ denotes element-wise vector multiplication. From this property, the kernels satisfy:

$$p_{n_i}^{\|\|^2}[n/L] = \hat{Q}_L(\hat{\mathbf{q}}_{n_i,L}^{row})^* = p_{n_i-1}^{\|\|^2}[((n - 1))_{NL}/L] \tag{4.53}$$

$$\vdots$$

$$= p_0^{\|\|^2}[((n - n_i))_{NL}/L] \tag{4.54}$$

From this relationship, all kernels can be obtained from circular shifts of a single basic kernel $p_0^{\|\|^2}[n/L]$ which will be denoted as $p^{\|\|^2}[n/L]$ or $\mathbf{p}^{\|\|^2}$.

The complete **SCR Algorithm** is described below:

- **Initialization.** Typically, these steps only need to be executed once. In many multicarrier applications, the transmitter of the communication system is not optimized for each specific channel such as broadcast DAB and DVB. For these cases, the initialization can be done off-line and the resulting parameters can be hard coded into the hardware of the transmitter. In other applications such as ADSL, the transmitter optimizes the modulation to the channel. For these cases, the PAR reduction algorithm will yield better performance if the initialization parameters are optimized to the channel during the training phase of the modem. The required steps are:

  1. Select the target PAR value or equivalently the desired level A. This will specify the required properties for the PRT set $\mathcal{R}$.

  2. Choose the set of reserved tones $\mathcal{R}$. The receiver must know that these tones are not used for data transmission. Strategies for generating $\mathcal{R}$ are described in Section 6.2 This step is very important since the convergence of this algorithm depends heavily on the PRT locations.

  3. Compute and store the kernel vector $\mathbf{p}^{\|_2}$ associated with $\mathcal{R}$. This kernel is based on a *2-norm* criteria. Section 6.1 describes alternative kernels based on other criteria.

- **Run time.** This algorithm is invoked for each multicarrier symbol that exceeds the desired PAR. For the *m-th* multicarrier symbol:

  1. Initial condition: set $\bar{\mathbf{x}}_L^{m(0)} = \mathbf{x}_L^m$

  2. Find the multicarrier symbol samples n_i for which $|\bar{x}_{n_i}^{m(i)}| > A$, i.e. the location of the large peaks. If all samples are below the target A, Jump to 5.

  3. Update $\bar{\mathbf{x}}_L^{m(i)}$ according to

  $$\bar{\mathbf{x}}_L^{m(i+1)} = \bar{\mathbf{x}}_L^{m(i)} - \mu \sum_{|\bar{x}_{n_i}^{m(i)}|>A} \alpha_{n_i}^{m(i)} p^{\|_2}[((n - n_i))_{NL}/L] \qquad (4.55)$$

  4. Increment the iteration counter, $i = i + 1$. If $i < MaxIterations$, Jump to 2.

  5. Transmit $\bar{\mathbf{x}}_L^{m(i)}$

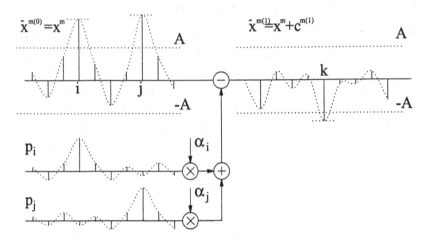

Figure 4.3: Illustration of the SCR gradient algorithm.

Figure 4.3 illustrates an example of the first step of the iterative SCR gradient algorithm. Realistic PAR targets yield values of A such that $Prob\{|\bar{x}_{n_i}^{m(i)}| > A\}$ is low and thus the number of terms in the summation in (4.55) is small. For each term in the summation, the number of operations is N compares, N additions and N multiplies. For complex (passband) multicarrier symbols, these operations are on complex numbers. Therefore, each step of the PAR reduction algorithm is achieved with sub-FFT complexity. To further simplify the number of computations per iteration, our simulations will restrict the summation in (4.55) only to its largest terms (largest values of $|\bar{x}_{n_i}^{m(i)}|$). This choice also allows for larger values for μ. Moreover, the N multiplications necessary to scale the kernel can be avoided all together by increasing the amount of memory allocated to the algorithm. Since $\mathbf{p}^{||2}$ is fixed, this is achieved by precomputing and storing $\mu_k \mathbf{p}^{||2}$ for a small set of values μ_k. The set of required values of μ_k can be reduced even further by noting that power of two scalings can be implemented very efficiently in hardware by left and right binary shifts. For example, if the transmitter precomputes and stores $\mathbf{p}^{||2}$ and $\sqrt{2}\mathbf{p}^{||2}$, all the scaled kernels of the form $2^{k/2}\mathbf{p}^{||2}$, for $k \in \mathcal{Z}$, require no multiplies[4]. In general, by storing the set of M vectors, $\mathbf{p}^{||2}$, $2^{1/M}\mathbf{p}^{||2}$, ..., $2^{(M-1)/M}\mathbf{p}^{||2}$, the transmitter can generate

---

[4]$\mathcal{Z}$ denotes the set of integers.

all the scaled kernels of the form $2^{k/M}\mathbf{p}^{\|2}$ for $k \in \mathcal{Z}$. Fortunately, the algorithm yields very good performance for small values of M.

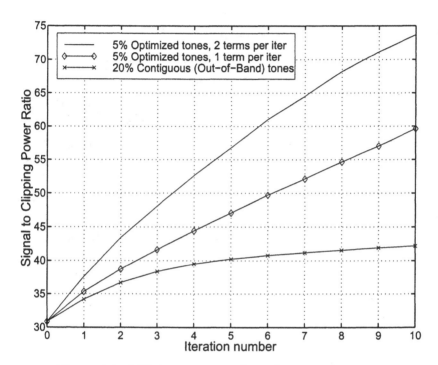

Figure 4.4: SCR improvement for a Structured tone set and a Randomly-Optimized set $\mathcal{R}^\star$ with SCR gradient technique.

Figure 4.4 plots the $SCR^{(i)}$ defined as:

$$SCR^{(i)} = \frac{\|\bar{\mathbf{x}}_L^{m(i)}\|_2^2}{\|\bar{\mathbf{x}}_L^{m(i)} - g_{sL}(\bar{\mathbf{x}}_L^{m(i)})\|_2^2} \qquad (4.56)$$

versus the number of iterations over the range of $i = 0, \ldots, 10$ for a real multicarrier transmitter with $N = 512$ and $ClipLevel = 9 \ dB$. Iteration 0 refers to the SCR of the original multicarrier symbol, before any PAR reduction, which in this case is 31 dB. The top-most curve refers to the an optimized PRT set $\mathcal{R}$ with $R/N = 5\%$ when the two largest terms in the summation of (4.55) are used at each iteration. For this example a 40 dB increase in the SCR is achieved after 10 iterations. The middle curve corresponds to using the same set $\mathcal{R}$ and only the largest term in the summation of (4.55). For this case, the convergence is slower but the complexity is lower. The bottom curve was generated by picking a set $\mathcal{R}$ comprised of contiguous tones with $R/N = 20\%$. Although the number of PRT has been increased by a factor of 4, the performance is very poor. This result shows the importance of choosing $\mathcal{R}$ correctly.

5.   ITERATIVE PAR REDUCTION AS A CONTROLLED CLIPPER

A second $\mathcal{O}(N)$ iterative algorithm can be constructed using the idea of a controlled clipper. Calling $g_{s_L}(\mathbf{x}_L^m) = \mathbf{x}_L^{m,clip}$ the clipped version of $\mathbf{x}^m$ for some given clipping level A, then $\mathbf{x}^{m,clip} = \mathbf{x}^m - \sum_{|x_{n_i}^m|>A} \beta_i \delta[(n - n_i)/L]$, where β_i are the clip complex values, and n_i the clip locations. By further calling $\mathbf{c}^{clip} = - \sum_{|x_{n_i}^m|>A} \beta_i \delta[(n - n_i)/L]$, the symbol $\mathbf{x}^{m,clip} = \mathbf{x}^m + \mathbf{c}^{clip}$ would have reduced the transmit symbol dynamic range to the interval $(-A, A)$. Therefore, $\mathbf{c}^{clip}$ can be thought of as a sum of delayed impulses scaled appropriately to cancel the peaks of $\mathbf{x}^m$. In general, for this choice of $\mathbf{c}^{clip}$, $\mathbf{C}^{clip} = DFT(\mathbf{c}^{clip})$ is nonzero over most frequencies, which distorts all the data vector components. On the other hand, if $\delta[n/L]$ can be approximated by a pulse $p[n/L] \approx \delta[n/L]$, then $\mathbf{c}^{clip}$ can be approximated as follows

$$\mathbf{c}^{appr} = - \sum_{|x_{n_i}^m|>A} \alpha_i p[((n - n_i))_{NL}/L] \approx - \sum_{|x_{n_i}^m|>A} \beta_i \delta[(n - n_i)/L] \quad (4.57)$$

where $p[((n - n_i))_{NL}/L]$ refers to the circular shift of $\mathbf{p}$ by n_i. For this scheme to work efficiently without distorting the data, two conditions are required:

1. The peak-reduction kernels $\mathbf{p} = p[n/L] = [p_0\ p_1 \cdots p_{NL-1}]$ should be as close as possible to the ideal impulse $\delta[n/L]$. Therefore, $\mathbf{p}$ must satisfy $p_0 = 1$, and $[p_L\ p_{L+1} \cdots p_{NL-1}]$ must be as *small* as possible.

2. The vector $\mathbf{C}^{appr} = DFT(\mathbf{c}^{appr})$ should be nonzero over a small set of preassigned frequencies. Thus, $\mathbf{C}^{appr} \in \mathcal{S}(\mathcal{R})$ where $\mathcal{S}(\mathcal{R})$ is the subspace of vectors with nonzero components in $\mathcal{R}$. Moreover the number of PRT must satisfy $R << N$.

Methods for generating peak-reduction kernels that satisfy the first condition are described in Section 6.1 The following analysis, which is based on simple properties of the DFT, shows that the second condition is met as long as $DFT(\mathbf{p}) = \mathbf{P} \in \mathcal{S}(\mathcal{R})$.

Let's assume the frequency kernel $\mathbf{P} \in \mathcal{S}(\mathcal{R})$. Then,

$$p[n/L] \overset{DFT}{\Longrightarrow} P_k = \begin{cases} P_{i_r}, & k = i_r \in \mathcal{R} \\ 0, & k \notin \mathcal{R} \end{cases} \tag{4.58}$$

From the shift property of the DFT operator,

$$p[((n-m))_{NL}/L] \overset{DFT}{\Longrightarrow} P_k e^{-j\frac{2\pi km}{NL}} = \begin{cases} P_{i_r} e^{-j\frac{2\pi i_r m}{NL}}, & k = i_r \in \mathcal{R} \\ 0, & k \notin \mathcal{R} \end{cases} \tag{4.59}$$

for $n, k, m \in \{0, 1, \ldots, NL-1\}$. Finally, using the linearity of the DFT,

$$\sum_i \alpha_i p[((n-n_i))_{NL}/L] \overset{DFT}{\Longrightarrow}$$

$$\sum_i \alpha_i P_k e^{-j\frac{2\pi kn_i}{NL}} = \begin{cases} \sum_i \alpha_i P_{i_r} e^{-j\frac{2\pi i_r n_i}{NL}}, & k = i_r \in \mathcal{R} \\ 0, & k \notin \mathcal{R} \end{cases} \tag{4.60}$$

Therefore, if $\mathbf{P} \in \mathcal{S}(\mathcal{R})$, then $\sum_i \alpha_i P_k e^{-j(2\pi k/NL)n_i} \in \mathcal{S}(\mathcal{R})$. Thus if $\mathbf{c}^{appr}$ can be expressed as $\mathbf{c}^{appr} = -\sum_i \alpha_i p[((n-n_i))_{NL}/L]$, then $DFT(\mathbf{c}^{appr}) = \mathbf{C}^{appr} \in \mathcal{S}(\mathcal{R})$ will also have only R nonzero values.

The next step for solving $\mathbf{c}^{appr}$ is to compute the values α_i and n_i. Attempting to optimize all α_i and n_i values simultaneously will lead to an LP for the real multicarrier case or a QCQP for the complex multicarrier case. Instead, an iterative algorithm similar to the **SCR Algorithm** described on page 81 can be used to compute the values of α_i and n_i. Using the **SCR Algorithm**, $\mathbf{c}^{appr}$ is a linear combination of the circularly shifted $\mathbf{p}$ as desired. Therefore, the **SCR Algorithm** can be interpreted as a special case of a controlled clipper where the general kernel $\mathbf{p}$ is replaced by $\mathbf{p}^{\|_2}$. Thus, the controlled clipper algorithm complexity remains $\mathcal{O}(N)$.

6. TONE RESERVATION KERNEL DESIGN

As described in the iterative algorithms in Sections 4. and 5., the PAR reduction signal $\mathbf{c}_L^{m(i)}$ must be computed for each data symbol $\mathbf{x}_L^m$ that exceeds our target PAR. Fortunately, the PRT set $\mathcal{R}$ and the kernel $\mathbf{p}$ are fixed during initialization and reused at each iteration for all symbols.

Since $\mathcal{R}$ and $\mathbf{p}$ only need to be calculated once, extra effort should be put on the kernel optimization. Moreover, the performance of the Tone Reservation algorithms depends heavily on the kernel properties. For example consider the simple case where $\mathcal{R}$ includes all the odd tones and $L = 1$, i.e.

$$
\begin{aligned}
x^m[n] + c^m[n] &= \frac{1}{\sqrt{N}} \sum_{\substack{k=0 \\ k \text{ even}}}^{N-1} X_k^m e^{j2\pi kn/N} + \frac{1}{\sqrt{N}} \sum_{\substack{k=0 \\ k \text{ odd}}}^{N-1} C_k^m e^{j2\pi kn/N} \\
&= \frac{1}{\sqrt{N}} \sum_{k=0}^{N/2-1} X_{2k}^m e^{j2\pi 2kn/N} + \frac{1}{\sqrt{N}} \sum_{k=0}^{N/2-1} C_{2k+1}^m e^{j2\pi(2k+1)n/N} \qquad (4.61) \\
&= \frac{1}{\sqrt{N}} \sum_{k=0}^{N/2-1} X_{2k}^m e^{j2\pi kn/(\frac{N}{2})} + \frac{1}{\sqrt{N}} \sum_{k=0}^{N/2-1} C_{2k+1}^m e^{j2\pi kn/(\frac{N}{2})} e^{j2\pi n/(\frac{N}{2})} \qquad (4.62)
\end{aligned}
$$

From (4.62), it is easy to show that $x^m[n] = x^m[n + N/2]$ and $c^m[n] = -c^m[n + N/2]$. Thus if $c^m[n]$ reduces the peak power at sample n, i.e. $|x^m[n] + c^m[n]| < |x^m[n]|$, this implies an increase in the peak power at the instant of $n + N/2$, i.e., $|x^m[n + N/2] + c^m[n + N/2]| = |x^m[n] - c^m[n]| > |x^m[n]|$. Thus, for this choice of $\mathcal{R}$, the optimal PAR reduction signal is $c^m[n] = 0$ and no PAR reduction can be achieved as $PAR\{x^m[n] + c^{m,opt}[n]\} = PAR\{x^m[n]\}$.

The importance of good kernel design can be also observed in Figure 4.3. In this figure, the symbol $\bar{x}^{m(0)}$ exceeded the *ClipLevel* at instants i and j. Although the kernel reduced the peak power at these two instants after the first iteration, a third peak appeared at instant k of $\bar{x}^{m(1)}$ due to the kernel side-lobes. Thus, minimization of kernel side-lobes is the design criteria for choosing the PRT locations.

6.1   COMPUTING PEAK REDUCTION KERNELS

For the $L = 1$ case, the kernel $\mathbf{p}$ should be a discrete-time impulse, i.e. $\mathbf{p} = [1 \ 0 \cdots 0]^T = \mathbf{e_0}$, for optimal performance. This way, every time the algorithm cancels a peak of $\bar{x}^{m(i)}$, no secondary peaks are generated at other locations. The problem with this choice is that $\mathbf{P}$ will have $R=N$ nonzero values. Therefore, the kernel should be designed to be as *close* as possible to $\mathbf{e_0}$ and still satisfy $R << N$. There are several solutions to $\mathbf{p}$ depending on the cost function chosen for computing the similarity between $\mathbf{p}$ and $\mathbf{e_0}$, i.e. $d(\mathbf{p}, \mathbf{e_0})$. The most natural and tractable solutions are those where $d(,)$ is a norm, i.e. $d(\mathbf{x}, \mathbf{y}) = \|\mathbf{x} - \mathbf{y}\|_l$. The case $l = 2$ is the Mean Square Error (MSE) solution which has a simple closed form. For the cases of $l = 1$ and $l = \infty$, the kernel can be solved via a LP for the real multicarrier case, and via a QCQP for the complex multicarrier case [Boyd and Vandenberghe, 1997].

Calling $\hat{\mathbf{P}}$ the nonzero values of $\mathbf{P}$, i.e. $\hat{\mathbf{P}} = [P_{i_0} \cdots P_{i_{R-1}}]^T$, a simple way to choose $\mathbf{p} = \hat{Q}_L \hat{\mathbf{P}}$ is using the MSE criterion, i.e.

$$\hat{\mathbf{P}}^{\|_{2,b}} = \arg \min_{\hat{\mathbf{P}}} \|\hat{Q}_L \hat{\mathbf{P}} - \mathbf{e}_0\|_2 \tag{4.63}$$

and the MSE time-domain kernel is

$$\mathbf{p}^{\|_{2,b}} = \hat{Q}_L \hat{\mathbf{P}}^{\|_{2,b}} \tag{4.64}$$

The solution for this case has a simple closed form given by,

$$\hat{\mathbf{P}}^{\|_{2,b}} = (\hat{Q}_L^* \hat{Q}_L)^{-1} \hat{Q}_L^* \mathbf{e}_0 = \frac{1}{L} \hat{Q}_L^* \mathbf{e}_0 = \frac{1}{L\sqrt{N}} \mathbf{1}_R \tag{4.65}$$

Thus, the MSE kernel is

$$\mathbf{p}^{\|_{2,b}} = \frac{1}{L\sqrt{N}} \hat{Q}_L \mathbf{1}_R \tag{4.66}$$

From (4.66), the MSE solution given by $\mathbf{p}^{\|_{2,b}}, \hat{\mathbf{P}}^{\|_{2,b}}$ is biased, since $p_0 = \frac{R}{NL} < 1$. The condition $p_0 = 1$, can be obtained by scaling the kernel:

$$\hat{\mathbf{P}}^{\|_2} = \frac{\sqrt{N}}{R} \mathbf{1}_R \tag{4.67}$$

and

$$\mathbf{p}^{\|_2} = \frac{\sqrt{N}}{R} \hat{Q}_L \mathbf{1}_R \tag{4.68}$$

From (4.68), the MSE kernel $\mathbf{p}^{\|_2}$ has a straightforward closed form solution. Solving (4.68) directly requires NR multiplies and adds. Since $\hat{Q}_L$ is a submatrix of the IDFT matrix of size NL scaled by $\sqrt{L}$, it can also be computed using an NL point IFFT with the PRT tones set to one.

Alternatively, an $\infty-norm$ constraint $(l = \infty)$ can be used to minimize the largest secondary peak of the kernel $\mathbf{p}$, i.e.,

$$\hat{\mathbf{P}}^{\|\infty} \quad = \quad \arg\min_{\hat{\mathbf{P}}} \|[p_L \; p_{L+1} \cdots p_{(N-1)L}]\|_\infty \qquad (4.69)$$

$$\text{subject to}: \quad \begin{array}{l} p_0 = 1 \\ \mathbf{p} = \hat{Q}_L \hat{\mathbf{P}} \end{array} \qquad (4.70)$$

As mentioned earlier, the solution to this problem is a LP for the real multicarrier case and a QCQP for the complex case. Although this kernel produces slightly better convergence than (4.68), it is also harder to compute which probably does not justify the extra complexity.

As can be seen from (4.64) and (4.70), $\mathbf{p}^{\|2}$ and $\mathbf{p}^{\|\infty}$ only depend on $\mathcal{R}$. Therefore, the kernel only needs to be computed if there is a change of peak reduction tone locations. Moreover, if the set $\mathcal{R}$ is known *a priori*, the complexity involved in computing $\mathbf{p}$ is not important since we can carefully design $\mathbf{p}$ off line. In any case $\mathbf{p}^{\|2}$ is simple to compute and can be optimized during the initialization phase of a channel optimized transmitter.

6.2   CHOOSING THE PRT SET

Section 6.1 described several methods to compute $\mathbf{p}^{\|l}$ when the PRT set $\mathcal{R}$ is known. This section describes efficient algorithms to generate $\mathcal{R}$.

Given the above discussion, the natural choice is to select $\mathcal{R}$ such that the secondary peaks or sidelobes in $\mathbf{p}^{\|l}$ are minimized, i.e.:

$$\mathcal{R}^{\infty,opt} \quad = \quad \arg\min_{\{i_0,\dots,i_{R-1}\}} \|[p_L \; p_{L+1} \cdots p_{(N-1)L}]\|_\infty \qquad (4.71)$$

where $[p_0 \; \cdots \; p_L \; p_{L+1} \; \cdots \; p_{NL-1}]^T = \mathbf{p}^{\|l} = \hat{Q}_L \hat{\mathbf{P}}^{\|l}$. The time-domain kernel $\mathbf{p}^{\|l}$ can be replaced by $\mathbf{p}^{\|2}$, the solution to (4.68), or $\mathbf{p}^{\|\infty}$ the solution to (4.70) for a given index choice $\mathcal{R}$. This problem is NP-hard since the kernel $\mathbf{p}^{\|l}$ must be optimized over all possible discrete sets $\mathcal{R}$ and cannot be solved for practical values of N. Alternatively, very good results can be obtained by generating random sets $\mathcal{R}^1,\dots,\mathcal{R}^M$ and selecting the best, which is denoted $\mathcal{R}^\star$. This method for PRT selection will be called random set optimization. The importance of a good choice of $\mathcal{R}$ can be seen in Figure 4.5 for a real multicarrier system with $N = 512$. The dashed line plots the optimized PAR CCDF for the choice $\mathcal{R}_{1/2} = \{244, 245, \dots 256\}$ (contiguous reserved tones). For this choice of $\mathcal{R}$, the 10^{-5} clipping rate threshold is reduced from a PAR_0 of 15 dB to a PAR_0 of 11.6 dB, that is a reduction of 3.4 dB. Similar results were obtained for other very structured Tone Reservation sets, such as $\mathcal{R}_{1/2} = \{11, 31, \ldots 251\}$ (equally spaced tones). However, an additional gain can be obtained using the random set optimization method as shown by the open-diamond marked solid line. For this choice, the 10^{-5} clipping rate threshold is reduced from a PAR_0 of 15 dB to a PAR_0 of 8.8 dB, giving a reduction of 6.2 dB.

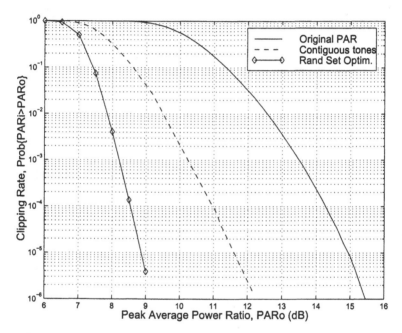

Figure 4.5: CCDF of $PAR\{x[n] + c[n]\}$ when $N = 512$
and $\frac{R}{N} = 5\%$ for two index choices,
Contiguous tones, $\mathcal{R}_{1/2} = \{244, 245, \ldots 256\}$
and Randomly-Optimized set $\mathcal{R}^*$.

The analysis above has focused on computing the set $\mathcal{R}$ of size R that minimizes the kernel side-lobes. For OFDM applications, where all tones have equal number of bits, this is the right problem formulation and the DRL is fixed since $DRL = TRR = R/N$. On the other hand, if the number of bits per tone is different, the DRL can be reduced by reserving the tones with low number of bits for PAR reduction as described in (4.15). For these applications, instead of minimizing the kernel side-lobes for a fixed number R of PRT, the kernel should be optimized over a given DRL:

$$\mathcal{R} = \left\{ \{i_0, \ldots, i_{R-1}\} \Big| \sum_{r=0}^{R-1} b_{i_r} \leq DRL \sum_{k=0}^{N-1} b_k \right\} \qquad (4.72)$$

Generating sets with the above constraint is straightforward. First, all tones for which $b_k = 0$ should be included in $\mathcal{R}$ since they do not reduce the data rate. Consider Tone Reservation PAR reduction with a given set $\mathcal{R}^{\alpha}$. If $b_{i_0} = 0$, and $i_0 \notin \mathcal{R}$, then the increased PRT set $\mathcal{R} = \mathcal{R}^{\alpha} \cup i_0$ will result in larger PAR reductions than the original set $\mathcal{R}^{\alpha}$ since the transmitter has extra degrees of freedom when solving (4.18) and (4.30). This analysis also applies if a power constraint is added to the PAR reduction signal $\mathbf{C}$ in (4.18) and (4.30). After all $b_k = 0$ tones are included, additional tones can be added randomly to the set until the DRL condition is met. The full optimization problem is to solve for (4.71) over all the PRT sets verifying the DRL constraint in (4.72). As mentioned above, a low complexity, approximate solution to (4.71) is random set optimization. For faster convergence, tones with less bits should be given higher probability in the random drawing, since they incur lower penalty in terms of DRL.

For channels where the number of bits per tone varies significantly, typically $DRL << TRR << 1$. On the other hand, if the $\{b_k\}$ values are very similar $DRL \approx TRR = R/N$. As described in the results, $R/N = 5\%$ yields up to 6 dB PAR reduction. If the 5% data reduction is considered too large, this value can be reduced significantly if our application can accommodate variable rate transmission or variable code rates. This can be done if the PAR reduction is activated only if necessary, i.e. if the $PAR\{\mathbf{x}^m\}$ exceeds some desired value, otherwise these tones are used for data transmission. For this case,

$$DRL = \frac{\sum_{r=0}^{R-1} b_{i_r}}{\sum_{k=0}^{N-1} b_k} Prob\{PAR\{\mathbf{x}^m\} > PAR_{desired}\} \qquad (4.73)$$

Assuming $b_k = const$, which is the worst case, then $DRL = TRR = \frac{R}{N} Prob\{PAR_m > PAR_{desired}\}$. One drawback of this technique is that the data rate is no longer constant. The multicarrier symbols for which $PAR_m < PAR_{desired}$ will transmit $\sum_{k=0}^{N-1} b_k$ bits, and the remaining multicarrier symbols will transmit $\sum_{k=0}^{N-1} b_k - \sum_{r=0}^{R-1} b_{i_r}$ bits. If constant data rate is required, then the transmitter can guarantee a fixed rate of $\sum_{k=0}^{N-1} b_k - \sum_{r=0}^{R-1} b_{i_r}$. The extra $\sum_{r=0}^{R-1} b_{i_r}$ bits can be used for additional redundancy or coding, for example to transmit extra bits from a punctured convolutional code. In any case, the number of bits reserved is small, and for most applications, is negligible.

6.3 NUMERICAL COMPUTATION OF PRT AND KERNEL

Sections 6.1 and 6.2 described several methods to compute the PRT locations and the peak reduction kernel. For multicarrier transmission without bit loading and static channels, these parameters should be computed off line and hard-coded into the modems. For multicarrier modulators with bit loading, better performance (larger PAR reductions or lower DRL) is achieved if these optimizations are calculated during initialization to take advantage of unequal b_k. This way, the kernel can be optimized for each channel. This section describes in more details how to compute $\mathcal{R}$ and $\mathbf{p}$ for the simplest method, the MSE kernel, and shows that it is very efficient to implement and therefore can be easily computed during initialization to achieve maximum performance. Since the kernel only needs to be computed once, the improved performance can justify the extra complexity needed. If the channel is quasi-static, then $\mathcal{R}$ and $\mathbf{p}$ can be updated periodically.

The following algorithm outlines the steps for the MSE kernel computation in (4.67) and (4.68):

1. Choose R random locations from the tone set $\{0, \ldots, N-1\}$ that verify (4.72), and call this set of indexes $\mathcal{R} = \{i_0, \ldots, i_{R-1}\}$.

2. Set $P[k] = \sqrt{N}/R$, for $k \in \mathcal{R}$, and $P[k] = 0$, for $k \notin \mathcal{R}$ (from (4.67)). That is, set the R reserved tones to $\sqrt{N}/R$, and all the remaining $N - R$ values to zero.

3. Compute $p[n/L] = IFFT(\sqrt{L}P[k])$ (from (4.68)). The largest value will be at p[0].

4. Compute the secondary peak of $p[n/L]$, that is $max\{|p[n/L]|\}$ for $n \in \{L, \ldots, (N-1)L\}$.

Step 1 requires generating R discrete uniform random values (a few more if there are some repeated indexes) and Step 2 requires N memory assignments. The complexity for the IFFT in Step 3 is about $4(NL/2)\log_2 NL$

operations[5]. Computing the secondary peak requires NL norm operations and NL compares. Thus, the total complexity to get a MSE peak reduction kernel is $R + N + 4(NL/2) \log_2 NL + NL + NL$ which is approximately the complexity of an NL point FFT.

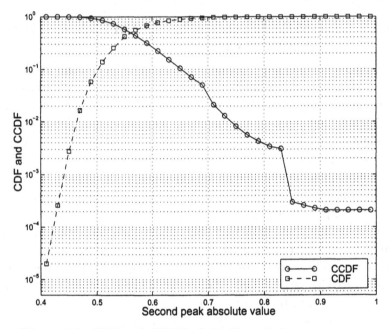

Figure 4.6: CDF and CCDF of the Kernel's largest
sidelobe for $R = 26$ and $N = 512$.

Since the PRT selection is based on random set generation, the following analysis evaluates the number of times the algorithm should be executed to obtain a good PRT set with very high confidence. The CCDF of the kernel second peak is defined as

$$CCDF(\beta) = Prob\{\text{kernel-sidelobes} > \beta\} \qquad (4.74)$$

Then, the probability that M randomly generated PRT kernels exceed β is given by $[CCDF(\beta)]^M$. If the kernel must satisfy *sidelobe* $< \beta$ with

---

[5]This estimate is for the complex multicarrier case. For the real case these numbers are smaller.

99.9999% confidence, M must be chosen to satisfy:

$$M \geq \frac{\log(10^{-6})}{\log(CCDF(\beta))} \qquad (4.75)$$

β	$CCDF(\beta)$	Number Iterations for 99.9999%	Calculation Time (millisec)	PAR Loss (dB)
.7	.97	4	2	.2
.65	.9	6	3	.1
.6	.72	11	6	$\approx .1$
.55	.4	28	14	$< .1$
.5	.09	147	74	$<< .1$

Table 4.1: PRT and kernel complexity.

In Figure 4.6 are plots of the side-lobe CDF and CCDF for the real multicarrier case with $R = 26$ and $N = 512$. Table 4.1 contains computed M for a few values of β. Assuming the transmitter has a multicarrier symbol rate[6] of $4KHz$, it must be able to compute 4000 IFFTs of size N a second. Using this as the processing power benchmark (this is a lower bound) Table 4.1 is a tabulation of the maximum time it takes to get the desired kernel with a 99.9999% confidence. For other levels of confidence, (4.75) can be easily modified. The last column in the table displays the PAR loss with respect to the best kernel found. As seen from the table, good kernels are generated after 14 milliseconds (less than .1 dB from the best one).

7.   RESULTS

Figure 4.7 and Figure 4.8 show the PAR CCDF improvement from applying the low complexity, iterative algorithm described in Section 5. with Nyquist sampling ($L = 1$). The oversampled cases ($L > 1$) are described in Chapter 5. For both figures, $N = 512$, $R/N = 5\%$ and the peak reduction kernel minimizes the 2-norm criteria. Figure 4.7 was generated with the index set $\mathcal{R}_{1/2} = \{244, 245, \ldots 256\}$. The index set for Figure 4.8 was computed using the random set optimization procedure described in Section 6.2 In both figures, the rightmost curve is the plot for the distribution of $PAR\{\mathbf{x}^m\}$. The second curve starting from the right is for $PAR\{\mathbf{x}^m + \mathbf{c}^{m(1)}\}$, i.e., after applying a single iteration. The third starting from the right is for $PAR\{\mathbf{x}^m + \mathbf{c}^{m(2)}\}$, i.e., after applying

---

[6]This is the symbol rate of ADSL modems.

two iterations and so forth. At each iteration, only the largest term in
the gradient update equation in (4.55) was used.

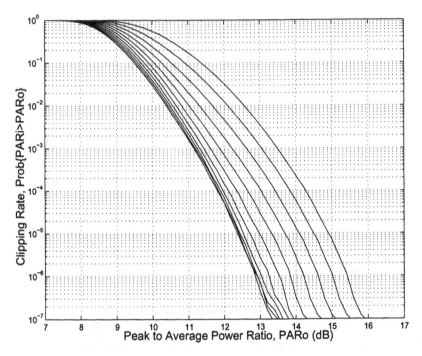

Figure 4.7: $PAR\{\mathbf{x}^m + \mathbf{c}^{m(i)}\}$ distribution for
$i = 1, \ldots, 10$ and $\frac{R}{N} = 5\%$. Contiguous tone
set with $\mathcal{R}_{1/2} = \{244, 245, \ldots 256\}$.

Assuming a desired clipping probability of 10^{-6}, from Figure 4.7,
about 0.4 dB of PAR reduction is achieved after one iteration, 0.8 dB
after two iterations, and 2 dB after 6 iterations. As can be seen by
comparing Figure 4.7 and Figure 4.8, the PAR reduction per iteration is
larger for the randomly optimized set $\mathcal{R}^\star$. For the same desired clipping
probability, the PAR reduction is about 2 dB after one iteration, 3 dB
for two iterations, and 4.5 dB after 6 iterations.

As mentioned in Section 2., adding $\mathbf{c}$ to the original symbol $\mathbf{x}^m$ increases the transmit power, ΔE. For the iterative methods, $\Delta E^{(i)}$ denotes the average power increase after i iterations. For the cases in
Figure 4.7 and Figure 4.8, after 10 iterations, $\Delta E^{(10)} = 0.13$ dB, which

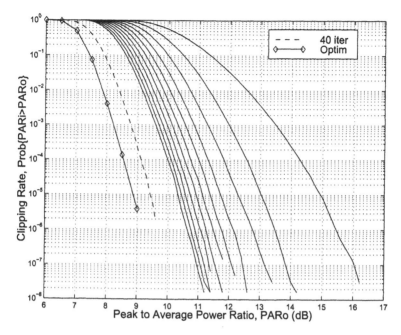

Figure 4.8: $PAR\{\mathbf{x}^m + \mathbf{c}^{m(i)}\}$ distribution for
$i = 1, \ldots, 10, 40$ and $\frac{R}{N} = 5\%$ with
Randomly Optimized set $\mathcal{R}^\star$.

is equivalent to a 3% power increase. This value is smaller than its counterparts in Section 2., where $\mathbf{c}^{m,opt}$ was computed exactly.

As described in Section 6.2, the DRL can be further reduced if variable bit rate or adaptive redundancy is allowed. From Figure 4.8, for a PAR target of $PAR_{desired} = 12\ dB$, a clipping probability of less than 10^{-7} can be achieved by correcting 3% of the symbols and with a maximum of 4 iterations. For these requirements, the complexity is $4NL$ and the $DRL = 5\% \times 3\% = .15\%$.

Chapter 5

PAR REDUCTION BY TONE INJECTION

C HAPTER 4 DESCRIBES THE FIRST ADDITIVE METHOD for reducing PAR in multicarrier transmission, and shows that reserving a small fraction of tones leads to large reductions in
PAR even with simple $\mathcal{O}(N)$ algorithms at the transmitter and with no
additional complexity at the receiver. When the number of tones N,
is small, the set of tones reserved for PAR reduction may represent a
non-negligible fraction of the available bandwidth and can result in a reduction in data rate, thus motivating the use of no-rate-loss PAR reduction methods. This chapter describes the second new additive method,
which achieves PAR reduction of multicarrier signals with no rate loss.
Most of the ideas presented here were first described in a number of
standards contributions [Tellado and Cioffi, 1998e, Tellado and Cioffi,
1998b, Tellado and Cioffi, 1998c] and in a patent application [Tellado
and Cioffi, 1997a], and more recently in [Tellado and Cioffi, 1998f]. All
these methods are based on the general additive model of PAR reduction described in Section 9. of Chapter 3, and more specifically on the
Tone Injection concept. The basic idea is to *increase* the constellation
size so that each of the points in the original basic constellation can be
mapped into several equivalent points in the expanded constellation. If
these duplicate signal points are spaced by $D = \rho d \sqrt{M}$, where M is
the constellation size, d is the distance between constellation points and
$\rho \geq 1$, the BER will not increase and the only addition to the standard
receiver is a *modulo-D* after the receiver Frequency Equalizer (FEQ).
Since each information unit can be mapped into several equivalent constellation points, these extra degrees of freedom can be exploited for

PAR reduction. The method is called *Tone Injection* as substituting the points in the basic constellations for the new points in the larger constellation is equivalent to *injecting* a tone of the appropriate frequency and phase in the multicarrier symbol. We would like to note some excellent simultaneous work by F. R. Kschischang, A. Narula and V. Eyuboglu. In [Kschischang et al., 1998a, Kschischang et al., 1998b], they independently proposed a PAR reduction method based on a lattice expansion of the original constellation, which corresponds to the special case $\rho = 1$ of our derivation.

Since the problem of choosing the constellation point within the equivalent set to minimize the PAR has exponential complexity, a suboptimal iterative algorithm with maximum complexity of $\alpha N^2 + \gamma N \sqrt{N}$, where $\alpha \ll 1$ and $\gamma \approx 1/(2\rho)$, is presented. Bounds on the maximum PAR reduction that can be achieved per Tone Injection step with the optimal solution are derived. It is shown that the iterative method is very close to this bound. It is also shown that choosing a larger separation between duplicate points leads to faster convergence and thus less complexity. For moderate values of N, the complexity of this Tone Injection iterative algorithm is low and comparable to the Tone Reservation method.

1.   PAR REDUCTION USING GENERALIZED CONSTELLATIONS

From (3.135), the additive PAR reduction methods can be formulated as:

$$\begin{aligned}
\bar{x}^m[n/L] &= x^m[n/L] + c^m[n/L] \\
&= \frac{1}{\sqrt{N}} \sum_{k=-\frac{N}{2}}^{\frac{N}{2}-1} (X_k^m + C_k^m)e^{j2\pi kn/NL} \quad (5.1)
\end{aligned}$$

Since the receiver must demodulate the received symbol $H_k(X_k^m + C_k^m) + N_k^m$, some method for removing C_k^m effectively is necessary for transparent PAR reduction. A number of structures of the PAR reduction vector $\mathbf{C}^m$ lead to simple receivers. In Chapter 4, the data tones and the reserved PAR reduction tones are different, and the separation is trivial. An alternative choice is $C_k^m = p_k^m D_k + j q_k^m D_k$ where the integers p_k^m and q_k^m are chosen to minimize the PAR, and the fixed constant D_k is known by the transmitter and receiver. By choosing D_k appropriately, C_k^m can be removed at the receiver by performing a modulo-D_k on the real and imaginary part of the output of the Frequency Equalizer (FEQ) $X_k^m + C_k^m + \tilde{N}_k^m = X_k^m + p_k^m D_k + j q_k^m D_k + \tilde{N}_k^m$. This structure is illustrated in Figure 5.1 and described in detail in the following discussion.

Assuming the k-th tone, X_k, has b_k bits, then X_k can have one of $M = 2^{b_k}$ discrete values from the MQAM constellation. For simplicity, let's assume a square constellation with a minimum distance between constellation points of d_k. Then, the real part of X_k, R_k and the imaginary part, I_k, can take values $\{\pm d_k/2, \pm 3d_k/2, \ldots, \pm(\sqrt{M_k} - 1)d_k/2\}$, where $\sqrt{M_k}$ is the number of levels per dimension. Figure 5.2 shows a $16QAM$ constellation, for which $b_k = 4$, and $\sqrt{M_k} = 4$.

Assume the k-th data element of the m-th symbol, X_k^m is the point given by $d_k/2 + j3d_k/2$ or $(1, 3)$ and denoted by A in Figure 5.2. By modifying the real and/or imaginary part of A, the PAR of the transmitted vector can potentially be reduced. To simplify the decoding step, A should be changed by an amount that can be estimated at the receiver. A very simple case is to transmit $\bar{A} = A + p_k D_k + j q_k D_k$ where p_k and q_k are any integer values and D_k is a positive real number known at the receiver. The most obvious choice, $p = q = 0$ gives us the original point, i.e. $\bar{A} = A$. The choice of $p_k = -1$ and $q_k = 0$ translates X_k^m from A to $\bar{X}_k^m = A - D = A2$.

To reduce the complexity of the receiver the value of D_k must be chosen carefully. For example, if $D_k = d_k \sqrt{M_k}/2$, $A2$ will overlap with B in Figure 5.2, and the receiver will confuse $\bar{A} = A - D = A2$ with B. Other choices, such as $D_k = d_k \sqrt{M_k}/2 + d_k/4$ will not overlap points in the transmitted constellation, but will reduce the minimum distance between possible transmit points to $d_k/4$ instead of the original d_k. On the other hand, by choosing $D_k \geq d_k \sqrt{M_k}$, the probability of decoding $\bar{A}$ erroneously at the receiver for an uncoded system is roughly the same as the probability of decoding A erroneously. Therefore, the SER will not change. The only extra complexity at the receiver is a modulo operation on the received complex vector. This operation requires subtracting/adding multiples of D_k from $\Re\{\bar{X}_k\} = R_k + p_k D_k$ and $\Im\{\bar{X}_k\} = I_k + q_k D_k$ until they both lie in the interval $[-D_k/2, D_k/2]$.

For example, generating all the values p_k and q_k over the range $|p_k| \leq 1$ and $|q_k| \leq 1$ for a 16QAM, span the new generalized constellation in Figure 5.3. This expanded constellation is equivalent to a $9 \times$ 16QAM constellation, where the horizontal and vertical translations between 16QAM sub-constellations is D_k. Therefore, the original 16QAM constellation points have been mapped to a bigger $9 \times$ 16QAM constellation

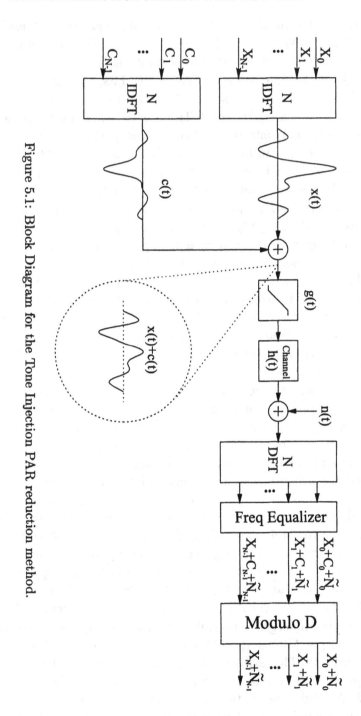

Figure 5.1: Block Diagram for the Tone Injection PAR reduction method.

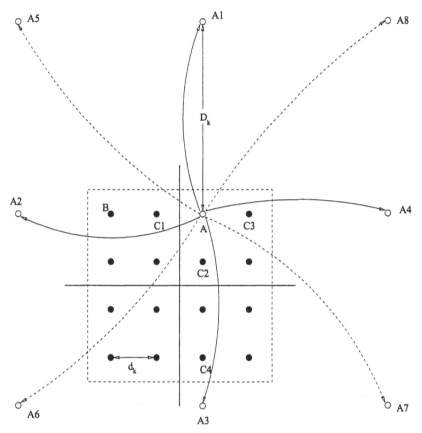

Figure 5.2: The constellation value A is the minimum
energy point of the equivalent set
$$\bar{A} = A + p_k D_k + j q_k D_k.$$

and the transmitter can choose from one of nine values to carry the
same information. These extra degrees of freedom can be used to generate multicarrier symbols with lower PAR. If the transmitter chooses
the special case where $D_k = d_k \sqrt{M_k}$, then the generalized constellation
in Figure 5.3 becomes a lattice [Kschischang et al., 1998a, Kschischang
et al., 1998b] (equally spaced points).

From Figure 5.2, it is clear that when $p \neq 0$ or $q \neq 0$, $\bar{A}$ has more
energy than A, and therefore the new multicarrier transmit symbol will
have more power. Fortunately, good choices of $\bar{A}$ can reduce the PAR
by over 5 dB with an average power increase of less than 2%. Figure
5.2 plots the nine values $\bar{A}$ with the lowest power for constellation point

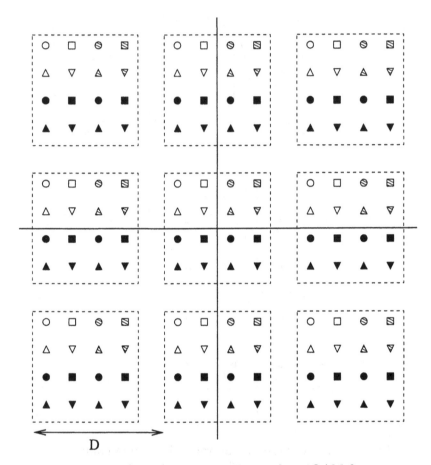

Figure 5.3: Generalized constellation for 16QAM for a given value D, when $|p_k| \leq 1$ and $|q_k| \leq 1$.

A and Figure 5.3 plots the nine equivalent values with the lowest power for each point in the standard 16QAM constellation.

As mentioned above, any choice of $D_k \geq d_k\sqrt{M_k}$ will not significantly increase the SER for an *uncoded system*, but the new nearest neighbors added due to the modulo operation can increase the probability of error for a coded system. For the standard 16QAM constellation in Figure 5.2, the nearest neighbors for A are $\{C1, C2, C3\}$. However, the nearest neighbors for the generalized constellation point $A3$ are $\{C1, C2, C3, C4\}$. Most codes are able to correct for the error $\{C1, C2, C3\}$, but probably not for $C4$. Thus, for coded systems, it may be beneficial to choose $D_k = \rho d_k \sqrt{M_k}$, where $\rho \geq 1$, to reduce the error in decoding $\{p, q\}$, i.e., the *modulo* error, to a desired value such that the new neighbor $C4$ does not become an issue. An additional benefit from choosing $D_k > d_k \sqrt{M_k}$ is that the transmitter can reduce the PAR with a smaller number of tone changes, as will be explained later. Other solutions are possible, such as not duplicating the outermost points of the original QAM constellation. If this algorithm only modifies the *interior points*, such as $C2$, the minimum distance for the *modulo* error is $2d_k$ instead of d_k, but this will increase the transmit power as described in Section 2. Another solution is to integrate the generalized constellation and the code. For example, with Trellis codes, the receiver needs to consider all the points in the generalized constellation when computing the metrics of the Viterbi decoder. This alternative will not increase the SER, but will increase the number of points in each coset and hence the receiver complexity.

Given these more general constellations, where each data sub-symbol X_k can be substituted for an equivalent $\bar{X}_k$, the transmitted multicarrier symbol can be chosen from the following set:

$$\bar{x}^m[n/L] = \frac{1}{\sqrt{N}} \sum_{k=-N/2}^{N/2-1} (X_k^m + p_k^m D_k + jq_k^m D_k) e^{j2\pi kn/NL}, \qquad (5.2)$$

where p_k^m, q_k^m are integers and $D_k \geq d_k \sqrt{M_k}$.

At the receiver end, decoding X_k^m from $\bar{X}_k^m$ is very simple. Since $D_k \geq d_k \sqrt{M_k}$, the receiver can decode X_k by performing a modulo operation, $X_k = mod_D\{\bar{X}_k\}$. By restricting $|p_k| \leq 1$ and $|q_k| \leq 1$, the receiver must perform at most $2N$ compares/adds per multicarrier vector symbol. Usually, the number of nonzero values for p_k, q_k is only a small fraction of N. Alternatively the modulo operation can be implemented very efficiently by appropriately scaling the decoded X_k^m values and masking the necessary bits.

2. POWER INCREASE

As described in the previous section, $\bar{X}_k = X_k + p_k D_k + jq_k D_k$ has more energy than X_k, whenever $p_k \neq 0$ or $q_k \neq 0$ and $D_k \geq d_k \sqrt{M_k}$. This section will show that the proper choices for p_k, q_k and the modified tone values X_k can greatly reduce this power increase. The importance of reducing this power increase is twofold. First, any power increase results in a reduction in SNR margin. Second, unnecessary power increases can lead to higher secondary peaks, which will complicate iterative greedy algorithms for computing $\bar{X}_k$. The 16QAM in Figure 5.2 will be used for the description, but MQAM constellations of any size can be studied following the same argument. Also, only the first *ring* of the generalized constellation is considered, i.e. the points for which $|p| \leq 1$ and $|q| \leq 1$. Further points can be used, but the energy increase for that tone will be very large, and thus these points probably will not be used in practice.

Consider first the point $A = (1, 3)$. When $d_k\sqrt{M_k} \leq D_k \leq 1.5d_k\sqrt{M_k}$, the set of equivalent points to A, sorted in ascending power is:

$$\{A, A3, A2, A6, A4, A7, A1, A5, A8\} \tag{5.3}$$

The power increase for these points is a function of the choice of D_k, and the constellation size M_k. For example if $\sqrt{M_k} = 4$ and $D_k = 1.25d_k\sqrt{M_k}$, $A3$ has 3 times more power than A, but $A8$ has 20 times the power in A. For the point $C2 = (1, 1)$, its minimum-power equivalent has 30 times the energy of $C2$. The reason for this is that $C2$ is close to the origin, and thus all its equivalent points are rather far away from the origin. Hence, if the target is to minimize the power increase, it's better to find equivalent values for tones with values close to the boundaries of the original constellation. That is, it makes more sense to modify X_k values equal to A or $C3$ but probably not $C2$.

The amount of energy in $x[n/L]$ from the k-th tone is $|X_k|^2$. The average energy of the k-th tone, $\mathcal{E}_k = E\{|X_k|^2\}$ satisfies:

$$\mathcal{E}_k = \frac{d_k^2}{6}(M_k - 1) \tag{5.4}$$

Thus, the minimum distance between constellation points is

$$d_k = \sqrt{\frac{6\mathcal{E}_k}{M_k - 1}} \tag{5.5}$$

In OFDM systems, all tones have equal energy and thus $\mathcal{E}_k = constant$. All the following derivations normalize the energy of each tone to unity, i.e., $\mathcal{E}_k = 1$. For DMT systems, bit loading is performed and the optimal energies per tone are not constant. Most of this variation is due to the granularity of the constellation size. For these cases, each tone is scaled by a gain g_k to keep the BER equal across all tones. Thus, the variation with respect to the mean power is denoted g_k^2. Assuming unity average tone power,

$$\mathcal{E}_k = g_k^2 \tag{5.6}$$

and

$$\sum_{k=0}^{N-1} \mathcal{E}_k = \sum_{k=0}^{N-1} g_k^2 = N \tag{5.7}$$

From (2.22) and (2.26), for constellation granularities of one bit (i.e. b_k are positive integers) and for a constant SNR gap (i.e. $\Gamma(\mathcal{C}, P_e) = const$) the energy scalings $\mathcal{E}_k = g_k^2$ are typically within the range $g_{min}^2 \leq g_k^2 \leq g_{max}^2$, where $g_{min} = -2\ dB$ and $g_{max} = 2\ dB$ [ANSI, 1995]. Since approximately a 3 dB increase (decrease) in SNR results in a 1 bit increase (decrease), a 1.5 dB energy adjustment will result in the closest integer bit. From (5.5) and (5.6), the distance can be written as

$$d_k = g_k \sqrt{\frac{6}{M_k - 1}} \tag{5.8}$$

where $g_k = 1$ for OFDM symbols.

Since the real part of X_k satisfies $R_k \in \{\pm d_k/2, \pm 3d_k/2, \dots, \pm(\sqrt{M_k} - 1)d_k/2\}$, the notation is simplified by introducing the variable r_k, where $R_k = r_k(d_k/2)$ and $r_k \in \{\pm 1, \pm 3, \dots, \pm(\sqrt{M_k} - 1)\}$. Similarly, the imaginary part can be written as $I_k = i_k(d_k/2)$. Since the tone power satisfies the additive property,

$$|\bar{X}_k|^2 = \bar{R}_k^2 + \bar{I}_k^2 = (R_k + p_k D_k)^2 + (I_k + q_k D_k)^2 \tag{5.9}$$

the power increase due to real and imaginary D-shifts can be studied independently. The following derivation confirms that any Tone Injection with $D_k \geq d_k \sqrt{M_k}$ increases the tone power, i.e.,

$$|\bar{X}_k|^2 = (R_k + p_k D_k)^2 + (I_k + q_k D_k)^2 > R_k^2 + I_k^2 = |X_k|^2 \tag{5.10}$$

when $|p_k| > 0$ or $|q_k| > 0$. Assuming $|p_k| > 0$ and $q_k = 0$, then,

$$\bar{R}_k^2 = \left(r_k \frac{d_k}{2} + \rho p_k d_k \sqrt{M_k} \right)^2 \tag{5.11}$$

$$= \left(\frac{d_k}{2} \right)^2 \left(r_k + 2\rho p_k \sqrt{M_k} \right)^2 \tag{5.12}$$

There are four cases to consider depending on the signs of p_k and r_k. When $p_k > 0$ and $r_k > 0$ then

$$\left(r_k + 2\rho p_k \sqrt{M_k} \right)^2 > r_k^2 \tag{5.13}$$

and (5.10) is satisfied. This is the case as well when both p_k and r_k are negative. The case where $p_k > 0$ and $r_k < 0$ satisfies,

$$\left(r_k + 2\rho p_k \sqrt{M_k} \right)^2 \geq \left(-(\sqrt{M_k} - 1) + 2\rho p_k \sqrt{M_k} \right)^2 \tag{5.14}$$

$$\geq \left(-(\sqrt{M_k} - 1) + 2\sqrt{M_k} \right)^2 \tag{5.15}$$

$$= \left(1 + \sqrt{M_k} \right)^2 \tag{5.16}$$

$$> r_k^2 \tag{5.17}$$

The remaining case where $p_k < 0$ and $r_k > 0$ is similar. This derivation also applies for the imaginary dimension, $|q_k| > 0$, and therefore D-shifts with $D_k \geq d_k \sqrt{M_k}$ always increase power.

The choices of constellation points R_k and shifts p_k that lead to minimum power increase can be evaluated as follows:

$$\arg \min_{r_k, p_k \neq 0} \left[\bar{R}_k^2 - R_k^2 \right] = \tag{5.18}$$

$$= \arg \min_{r_k, p_k \neq 0} \left(\frac{d_k}{2} \right)^2 \left[\left(r_k + 2\rho p_k \sqrt{M_k} \right)^2 - r_k^2 \right] \tag{5.19}$$

$$= \arg \min_{r_k, p_k \neq 0} \left[\left(r_k + 2\rho p_k \sqrt{M_k} \right)^2 - r_k^2 \right] \tag{5.20}$$

$$= \arg \min_{r_k, p_k \neq 0} \left[r_k^2 + 4 r_k \rho p_k \sqrt{M_k} + 4\rho^2 p_k^2 M_k - r_k^2 \right] \tag{5.21}$$

$$= \arg \min_{r_k, p_k \neq 0} \left[r_k p_k \sqrt{M_k} + \rho p_k^2 M_k \right] \tag{5.22}$$

$$= \arg \min_{r_k, |p_k| = 1} \left[r_k p_k \sqrt{M_k} + \rho M_k \right] \tag{5.23}$$

$$= \arg \min_{r_k, |p_k| = 1} \left[r_k p_k \sqrt{M_k} \right] \tag{5.24}$$

$$= \arg \min_{r_k, |p_k| = 1} r_k p_k \tag{5.25}$$

From (5.22) and (5.25), the minimum power increase occurs when $p_k = 1$ and $r_k = -(\sqrt{M_k} - 1)$ or when $p_k = -1$ and $r_k = \sqrt{M_k} - 1$. Thus, for the real part, the D-shift must be of opposite sign to the constellation real component, and the real component must have maximum amplitude. Moreover, a single D-shift must be performed. The equivalent argument applies for the imaginary part.

Substituting $p_k = 1$ and $r_k = -(\sqrt{M_k} - 1)$ in $\bar{R}_k^2 - R_k^2$, the minimum the power increase for a single D-shift is given by,

$$
\begin{aligned}
\min_{r_k, p_k \neq 0} \left[\bar{R}_k^2 - R_k^2\right] &= \left(\frac{d_k}{2}\right)^2 \left[\left(r_k + 2\rho p_k \sqrt{M_k}\right)^2 - r_k^2\right]_{p_k=1, r_k=-(\sqrt{M_k}-1)} \\
&= \left(\frac{d_k}{2}\right)^2 \left[r_k^2 + 4 r_k \rho \sqrt{M_k} + 4\rho^2 M_k - r_k^2\right]_{r_k=-(\sqrt{M_k}-1)} \\
&= \left(\frac{d_k}{2}\right)^2 \left[-4(\sqrt{M_k} - 1)\rho\sqrt{M_k} + 4\rho^2 M_k\right] \\
&= d_k^2 \rho \left[-(\sqrt{M_k} - 1)\sqrt{M_k} + \rho M_k\right] \\
&= d_k^2 \rho \left[\sqrt{M_k} + (\rho - 1)M_k\right] \\
&= \frac{6\rho g_k^2}{M_k - 1} \left[\sqrt{M_k} + (\rho - 1)M_k\right]
\end{aligned}
\tag{5.26}
$$

Since the average energy of the multicarrier symbol is given by (5.7), the minimum power increase relative to the the symbol power is given by

$$
\min_{r_k, p_k \neq 0} \left[\bar{R}_k^2 - R_k^2\right] = \frac{6\rho g_k^2}{N(M_k - 1)} \left[\sqrt{M_k} + (\rho - 1)M_k\right]
\tag{5.27}
$$

For the OFDM special case with minimum shift, $\rho = 1$, (5.27) simplifies to

$$
\min_{r_k, p_k \neq 0} \left[\bar{R}_k^2 - R_k^2\right] = \frac{6\sqrt{M_k}}{N(M_k - 1)}
\tag{5.28}
$$

From (5.28), as the constellation size grows, the minimum power increase for the outer points in the constellation becomes smaller. As described in Section 3., only a small fraction of tones need to be adjusted. Therefore, by selecting the shifts using the argument above, the total power increase of the multicarrier symbol can be below 2% on the average.

3.   MAXIMUM PAR REDUCTION PER DIMENSION TRANSLATION

Although the following sections focus on real multicarrier systems, these results can be extended easily to include the complex case. After quantifying the symbol power increase due to Tone Injection in Section 2., this section quantifies the maximum achievable PAR reduction from Tone Injection per dimension translation when the first set of constellation neighbors is used, i.e. $|p_k| \leq 1$ and $|q_k| \leq 1$.

For the real multicarrier case, assuming N is even, the transmitter IFFT output vector is[1]:

$$x[n/L] = \frac{2}{\sqrt{N}} \sum_{k=1}^{N/2-1} [R_k \cos(2\pi kn/NL) - I_k \sin(2\pi kn/NL)] \quad (5.29)$$

where $X_k = R_k + jI_k$. If the PAR of this multicarrier symbol is high, there must be at least one value of $|x[n/L]|$ that is large. Assuming the location of the maximum value is n_0, we can replace $n = n_0$ in (5.29):

$$x[n_0/L] = \frac{2}{\sqrt{N}} \sum_{k=1}^{N/2-1} [R_k \cos(2\pi kn_0/NL) - I_k \sin(2\pi kn_0/NL)] \quad (5.30)$$

Let's assume that $x[n_0] > 0$ and that $\cos(2\pi k_0 n_0/NL) = l > 0$ for some frequency bin k_0. If we subtract[2] D_{k_0} from R_{k_0} the new transmit multicarrier symbol vector $\bar{x}[n/L]$ can be computed without repeating the IFFT because the algorithm has only modified one tone. The new symbol is simply

$$\bar{x}[n/L] = x[n/L] + \frac{2}{\sqrt{N}}(-D_{k_0})\cos(2\pi k_0 n/NL) \quad (5.31)$$

After replacing R_{k_0} for $R_{k_0} - D_{k_0}$, the reduced peak at $\bar{x}[n_0]$ will satisfy the relationship:

$$\bar{x}[n_0/L] = x[n_0/L] - \frac{2lD_{k_0}}{\sqrt{N}} \quad (5.32)$$

$$= x[n_0/L] - \frac{2l\rho d_{k_0}\sqrt{M_{k_0}}}{\sqrt{N}} \quad (5.33)$$

$$= x[n_0/L] - \frac{2lg_{k_0}\rho}{\sqrt{N}}\sqrt{\frac{6M_{k_0}}{M_{k_0} - 1}} \quad (5.34)$$

---

[1]The DC term (R_0) and the Nyquist term ($R_{N/2}$) have been set to zero to simplify the discussion.
[2]The values of D_k can be different from tone to tone. Moreover different values of D_k for the real and imaginary part can be used.

Thus, the maximum peak reduction of sample $x[n_0/L]$ that can be achieved after shifting one dimension occurs when $l = 1$. The new peak will satisfy:

$$\bar{x}[n_0/L] \geq x[n_0/L] - \frac{2g_{k_0}\rho}{\sqrt{N}}\sqrt{\frac{6M_{k_0}}{M_{k_0} - 1}} \qquad (5.35)$$

Since other secondary peaks might appear at other locations, $n_1 \neq n_0$, the new multicarrier symbol will satisfy:

$$\max_n |\bar{x}[n/L]| \geq \max_n |x[n/L]| - \frac{2g_{k_0}\rho}{\sqrt{N}}\sqrt{\frac{6M_{k_0}}{M_{k_0} - 1}} \qquad (5.36)$$

A similar argument follows for all other permutations. If

$$\cos(2\pi k_0 n_0/NL) = l < 0, \qquad (5.37)$$

$R_{k_0} + D$ must be substituted for R_{k_0}. These ideas also can be extended to the $I_k \sin(2\pi k n_0/NL)$ terms. In general, the single dimension *D-shift* update (5.31), can take any of the following 4 options for each tone k_0,

$$\bar{x}[n/L] = x[n/L] + \frac{2}{\sqrt{N}}(\pm D_k)\{cos, sin\}(2\pi k_0 n/NL), \qquad (5.38)$$

where $n = 0, \ldots, NL - 1$. From (5.36), the maximum peak reduction per Tone Injection step is:

$$\delta_k = \frac{2g_k\rho}{\sqrt{N}}\sqrt{\frac{6M_k}{M_k - 1}} \qquad (5.39)$$

If this procedure is followed on K real/imaginary dimensions simultaneously, the maximum peak reduction is $\sum_{k=1}^{K} \delta_k$. From (5.39), the peak reduction factor δ_k decreases as N increases, which increases the number of iterations needed to reduce the PAR to the target value. For example, if g_k, M_k and ρ are constant for all tones (OFDM) and a peak reduction of Δ is desired, the number of iterations needed is

$$K = \frac{\Delta}{\delta} = \frac{\sqrt{M-1}}{2\rho\sqrt{6M}}\Delta\sqrt{N} = \gamma_\Delta\sqrt{N} \qquad (5.40)$$

Thus, for larger values of N, increasing ρ is a good choice since it reduces the number of iterations. Table 5.1 lists the maximum PAR reduction from multiple tone modifications for three different values of ρ. An IFFT size of 64 with 6 bits/tone is considered and the original symbol worst case PAR is 15 dB (clipping rate 10^{-7}).

Iterations	$\rho = 1$	$\rho = 1.125$	$\rho = 1.25$
1	1(dB)	1.1(dB)	1.3(dB)
2	2.2	2.5	2.8
3	3.5	4	4.6
4	5	6	7

Table 5.1: Maximum PAR reduction (in dB) vs. number of iterations for 64QAM and $N = 64$.

Table 5.2 assumes an IFFT size of 512 with 4bits/tone and an original symbol worst case PAR of 15.5 dB. This table shows that the case of $\rho = 1$ requires 50% more iterations/complexity than $\rho = 1.5$.

Iterations	$\rho = 1$	$\rho = 1.25$	$\rho = 1.5$
1	.3(dB)	.4(dB)	.5(dB)
2	.7	.9	1
3	1	1.3	1.6
4	1.4	1.8	2.2
6	2.2	2.9	3.6
8	3.1	4.1	5.2
10	4.1	5.5	7.2

Table 5.2: Maximum PAR reduction (in dB) vs. number of iterations for 16QAM and $N = 512$.

4.   SIMPLE ALGORITHMS FOR COMPUTING $\bar{X}[N/L]$

The previous section provided upper bounds on the PAR reduction that is possible from these generalized constellation methods. This section describes some simple algorithms that are close to achieving these upper bounds with small transmit power increase and low complexity.

Rewriting (5.2) for the baseband case produces,

$$\bar{x}[n/L] = \frac{2}{\sqrt{N}} \sum_{k=1}^{N/2-1} \Big[(R_k + p_k D_k) \cos(2\pi kn/NL)$$
$$- (I_k + q_k D_k) \sin(2\pi kn/NL) \Big] \quad (5.41)$$

Finding the values of p_k and q_k that produce the lowest PAR for $\bar{x}[n/L]$ requires solving an integer programming problem, which has exponential complexity. Assuming S equivalent points per basic constellation point, if K dimensions are to be D-shifted, there are

$$\binom{N}{K}S^K = \frac{N(N-1)\cdots(N-K+1)}{K!}S^K \qquad (5.42)$$

$$\approx \frac{N^K}{K!}S^K \qquad (5.43)$$

$$\approx (NS)^K \qquad (5.44)$$

combinations for the vectors $[p_0 \dots p_{\frac{N}{2}-1}]$ and $[q_0 \dots q_{\frac{N}{2}-1}]$. Since $K \ll N$, the simple approximations in (5.43) and (5.44) are reasonable. From the argument above, a peak reduction of amplitude Δ requires $K = \gamma_\Delta \sqrt{N}$ dimension shifts, where γ_Δ is given by (5.40). Therefore, the number of combinations required for a fixed PAR reduction is:

$$\binom{N}{\gamma_\Delta \sqrt{N}}S^{\gamma_\Delta \sqrt{N}} \approx (NS)^{\gamma_\Delta \sqrt{N}} \qquad (5.45)$$

which explains the exponential complexity of the optimal solution. For a PAR reduction of 5 dB, $M \geq 4$ and $\rho = 1$, $\gamma_\Delta \approx 1/2$. Fortunately, good approximations to the optimal solution are possible with low complexity with an iterative algorithm that utilizes the principles described in the previous sections.

First, a nonzero value for p_k or q_k increases the symbol energy. Although the algorithm should reduce the larger peaks it will inevitably increase other samples of the multicarrier symbol. To facilitate subsequent iterations, choosing the equivalent points that reduce the power increase is important. For example, if all tones transmit 16QAM constellations, choosing the terms where $|r_k| = 3$ or $|i_k| = 3$ minimizes the transmitter power increase by selecting $sign(p_k) = -sign(R_k)$ or $sign(q_k) = -sign(I_k)$. Second, choosing the tones where the sinusoid value, l in (5.32), is large at the peak locations, yields larger PAR reductions per step. Similarly, larger values of D will cancel the peak faster but at the expense of larger power increases.

The algorithm starts with the original multicarrier symbol ($p_k = 0$ and $q_k = 0$). After finding the location of the maximum, n_0, we find a tone k_0 such that either $|R_{k_0}|$ or $|I_{k_0}|$ is large and such that the sinusoid is of the opposite sign and close to one in magnitude at n_0. After the desired tone, k_0, is found, $\bar{x}[n/L]$ is updated using (5.38). If more than one value of $x[n/L]$ is large, we must find a tone k_0 that reduces as many peaks as possible. This procedure can be repeated several times until the desired PAR is achieved or the maximum number of iterations or maximum transmit power increase has been reached. This procedure was used to generate the figures in Section 5.

5.   RESULTS

Figures 5.4 and 5.5 compare the multicarrier symbol PAR CCDF at the Nyquist sampled IFFT output $(L = 1)$. The simulation parameters for Figures 5.4 are: DFT of size 64, 16QAM, $\rho = 1$ and the maximum number of single tone iterations, $K = 4$, or equivalently, a maximum of 4 values of R_k and I_k combined have changed. The PAR reduction is about 5 dB at a clipping rate of 10^{-6}, which is the maximum reduction for 4 iterations from Table 5.1. The average power increase is only 1.3%. Figure 5.4 also includes PAR results for a more complicated double-tone algorithm. This algorithm reduces the PAR by 6.5 dB at the expense of a larger power increase and an additional iteration of the algorithm.

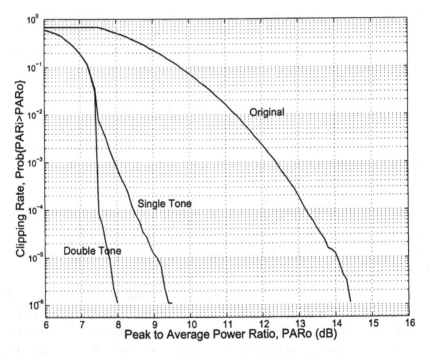

Figure 5.4: Tone Injection PAR CCDF for $N = 64$, 16QAM and $\rho = 1$.

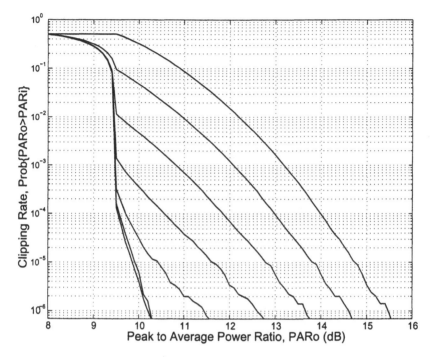

Figure 5.5: Tone Injection PAR CCDF for $N = 256$,
16QAM and $\rho = 1.75$ for iterations
$i = 1, \ldots, 6$.

Figure 5.5 compares the PAR reduction for the first 6 iterations for a
DFT of size 256, with 16QAM in each tone and $\rho = 1.75$. The proposed
algorithm reduces PAR by more than 5 dB at a clipping rate of 10^{-6}.
This is achieved with a maximum of 6 nonzero values of p_k, q_k, or equivalently, a maximum of 6 values of R_k and I_k combined have changed.
This is only .5 dB away from the maximum reduction achieved by the
exponentially hard algorithm as shown in Table 5.2. The average power
increase is only 1.6%

Figures 5.6-5.9 study the effect of oversampling and filtering on the
Tone Injection PAR reduction algorithms. In Section 4. of Chapter 3 the
performance of an ideal oversampled PAR reduction method after filtering was described. The ideal oversampled Tone Injection PAR reduction
algorithm would require PAR reduction on the oversampled IFFT output $\bar{x}[n/L]$, where $L > 1$. Lower complexity, approximate solutions
can be derived by oversampling (interpolating) the output of a critically

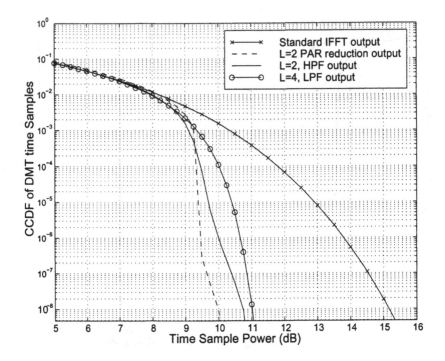

Figure 5.6: Sample CCDF at four different points of an
ADSL transmitter: standard IFFT output,
oversampled Tone Injection PAR reduction
output, oversampled FIR HPF output
($L = 2$), and oversampled FIR LPF output
($L = 4$).

sampled IFFT. The complexity of the oversampled algorithms increases
linearly with the oversampling factor. The derivations in Section 4. of
Chapter 3 showed that most of the PAR reduction is preserved after
ideal interpolation if PAR reduction was computed on a $L = 2$ oversampled signal. The simulations in this section confirm that most of
the Tone Injection PAR reduction on a $L = 2$ oversampled signal is still
preserved on the continuous-time Analog Front End (AFE) input after
practical filtering. In Figure 5.6, $L = 2$ was chosen, and the same FIR
filters described in Figure 3.12 were used. For this case the PAR at the
AFE input is reduced by more than 4 dB. In Figure 5.7 2× oversampled

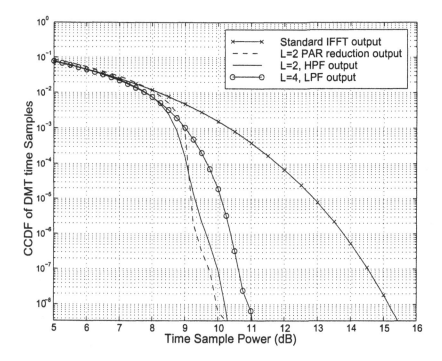

Figure 5.7: Same as Figure 5.6 with Butterworth HPF
and Butterworth LPF.

PAR reduction was performed and the analog LPF is a *4-th* order Butterworth filter. The filters used in Figure 5.8 were provided by Pairgain
and $L = 4$ PAR reduction was performed. For $L = 2$ PAR reduction the
final PAR is about 0.4 *dB* higher.

Moreover, if the filters are known, with some extra complexity the
effect of the filters can be included, and the PAR reduction algorithm
can be modified to reduce the PAR after filtering. This way the full 5-
6 *dB* reduction in PAR can be approached. Figure 5.9 plots the sample
CCDF at the AFE when the PAR reduction takes into account the effects
of the transmit filters. For this case the filters are the same FIR filters
used in Figure 5.6. Taking the filters into account reduces the PAR by
5 *dB*.

The main focus of the algorithms used in Figures 5.4-5.9 is to minimize
the transmit power increase with low complexity instead of achieving
the absolute minimum PAR. Larger reduction in PAR are possible by

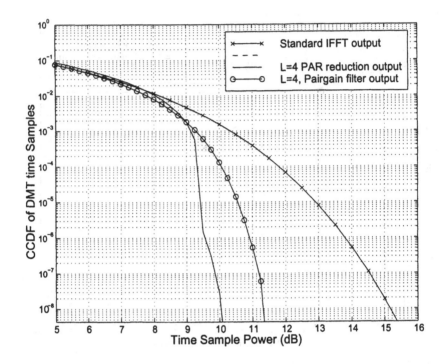

Figure 5.8: Same as Figure 5.6 for the ADSL transmit
filter provided by Pairgain.

relaxing the constraint on average power increase, increasing the number
of iterations or by using more complex algorithms.

6. CONCLUSIONS

An additive method for reducing PAR with no data rate loss has been
presented. This method is based on expanding the transmit constellation
by multiples of some fixed value. These generalized constellations can
help reshape the original time-domain multicarrier symbol and can easily
reduce its peaks by more than 4 *dB* with low complexity and negligible
transmit power increase. Moreover, these generalized constellations can
be easily mapped into the original constellation with a simple modulo
operation at the receiver. Depending on the application constraints,
this algorithm can be applied with different levels of complexity and
performance.

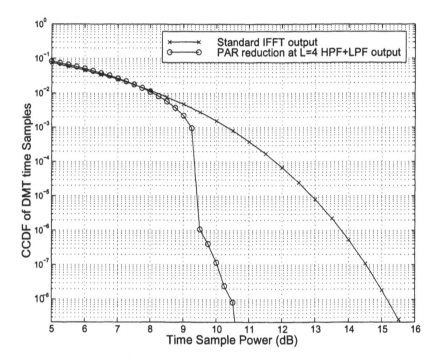

Figure 5.9: Peak CCDF at the 4× oversampled filtered
output when the transmit filters are included
in the PAR reduction algorithm.

Chapter 6

MAXIMUM LIKELIHOOD DETECTION OF DISTORTED MULTICARRIER SIGNALS

S INCE MULTICARRIER SIGNALS exhibit relatively high PAR, if linear operation is required over the full dynamic range, the average transmitter power must be reduced and this would degrade the received SNR. To avoid operating transceiver components with extremely large power reductions[1], the designer has two alternatives. The first alternative is to generate multicarrier signals with low PAR using some of the methods summarized in Chapter 3. Most of these methods add substantial complexity at the transmitter, require significant coding overhead or degrade the SNR of the system. The methods in Chapter 4 and Chapter 5 require small coding overhead and negligible SNR degradation but require some extra complexity (comparable to a FFT). Moreover, these methods still require the channel to be linear over the reduced peak-to-peak operating region. The second alternative for reducing the PAR, which is the simplest solution for the transmitter, allows saturation of the power amplifiers or clipping in the DAC/ADC, which leads to nonlinear distortion that cannot be corrected with standard linear receivers. There has been a number of publications that quantify the multicarrier signal degradation caused by nonlinearities. Some of the early work on this topic was done in [Mestdagh et al., 1994, Gross and Veeneman, 1994] for baseband DMT, and in [O'Neill and Lopes, 1994, O'Neill and Lopes, 1995] for passband OFDM. All these papers assume that the nonlinearity is an ideal limiter and compute the SNR degradation and

---

[1]The term used in practice is *power back-off*

the Power Spectral Density (PSD) of the clipped/limited signal without providing any methods to correct for the additive clipping noise. Certainly, these two alternatives can be used simultaneously to achieve better performance.

If the nonlinear characteristic of the transmitter is known, the nonlinear distortion is a deterministic function of the data that can be corrected at the receiver. Section 1. quantifies the loss in mutual information caused by the non-ideal properties of the transmitter and shows that this loss is relatively small. Section 2. formulates the optimal Maximum Likelihood (ML) receiver in the presence of nonlinear distortion and shows that ML has very high complexity. Moreover it proposes a suboptimal ML demodulator that iteratively estimates the nonlinear distortion and that has small SNR degradation for practical values of PAR. Therefore, the PAR of the multicarrier signal can be reduced at the transmitter by using a simple clipper or by allowing saturation of the DAC or amplifiers, and the non-linearly distorted signal can be recovered with very low degradation using the proposed iterative ML algorithm. The spectral regrowth introduced by the clipper can be significantly reduced using methods in [Pauli and Kuchenbecker, 1997, van Nee and de Wild, 1998, Pauli and Kuchenbecker, 1998]. Section 3. describes some numerical examples for the approximate ML receiver.

Most of the ideas presented here were first described in a patent application [Tellado and Cioffi, 1997a] and more recently in [Tellado and Cioffi, 1999a, Tellado and Cioffi, 1999b, Tellado and Cioffi, 2000a].

1.   MEMORYLESS NONLINEARITY EFFECTS ON ACHIEVABLE RATE

In this section, the loss of mutual information caused by a memoryless nonlinearity is evaluated for the AWGN discrete-time memoryless channel with independent Gaussian inputs. When the subchannel elements X_k^m are Gaussian i.i.d. random variables, and the CP has zero length, the Nyquist-sampled IDFT outputs $x[n]$ are i.i.d. Gaussian random process as described in Section 3. of Chapter 3. Thus, the model shown in Figure 6.1, where x is i.i.d. Gaussian, n is the AWGN, and y is the channel output is an accurate model when the nonlinearity acts on the Nyquist-sampled IDFT outputs $x[n]$ and the remaining transceiver elements are ideal. For simplicity, this analysis will focus on the case where the input signal x is real, but these derivations can be extended to complex signals.

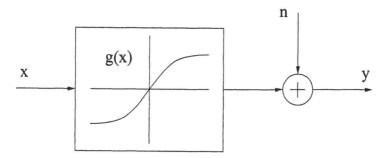

Figure 6.1: Channel nonlinear model for computing mutual information.

The mutual information, for the real valued case, can be written as

$$
\begin{aligned}
I(x;y) &= I(x;g(x)+n) &\quad (6.1)\\
&= h(g(x)+n) - h(g(x)+n|x) &\quad (6.2)\\
&= h(g(x)+n) - h(n) &\quad (6.3)\\
&= h(g(x)+n) - \frac{1}{2}\log_2(2\pi e\sigma_n^2) &\quad (6.4)
\end{aligned}
$$

In the formulae above, σ_n^2 is the variance of the Gaussian noise and $h(z)$ denotes the differential entropy of a random variable z and is defined to be [Cover and Thomas, 1991]

$$
h(z) = -\int p(z)\log_2 p(z)dz \qquad (6.5)
$$

where $p(z)$ is the probability density function for z.

The first term in (6.4) can be upper-bounded as follows:

$$
\begin{aligned}
h(g(x)+n) &\leq \frac{1}{2}\log_2(2\pi e(\sigma^2[g(x)]+\sigma_n^2)) &\quad (6.6)\\
&\leq \frac{1}{2}\log_2(2\pi e(\sigma_x^2+\sigma_n^2)) &\quad (6.7)
\end{aligned}
$$

where the first inequality follows from the independence of $g(x)$ and n and from the property that the Gaussian distribution maximizes the differential entropy for an average power limited random variable. The notation $\sigma^2[g(x)]$ is used to denote the variance of the random variable $g(x)$. The second inequality follows from the the non-expansive property of $g(\cdot)$.

$$\sigma^2[g(x)] \leq E\{g(x)^2\} \tag{6.8}$$

$$= \int g(x)^2 p_x(x) dx \tag{6.9}$$

$$\leq \int x^2 p_x(x) dx \tag{6.10}$$

$$= \sigma_x^2 \tag{6.11}$$

From (6.1)-(6.7), we thus have

$$I(x; y) \leq \frac{1}{2}\log_2(2\pi e(\sigma_x^2 + \sigma_n^2)) - \frac{1}{2}\log_2(2\pi e \sigma_n^2) \tag{6.12}$$

$$= \frac{1}{2}\log_2(1 + \frac{\sigma_x^2}{\sigma_n^2}) = C[g(x) = x] \tag{6.13}$$

where $C[g(x) = x]$ represents the capacity of the channel when the nonlinearity $g(x)$ is ideal i.e., linear. Therefore, (6.13) implies that the non-expansive nonlinearity $g(\cdot)$ decreases the mutual information of the channel.

So far, we have shown that non-expansive memoryless nonlinearities reduce mutual information for discrete AWGN channels. We will now compute $I(x; y)$ exactly for specific nonlinearities $g(x)$. Computing $I(x; y)$ for a given $g(x)$ requires knowledge of the pdf of $g(x) + n$. Since the Gaussian i.i.d. random variables x and n are independent, so will $w = g(x)$ and n. Thus, the pdf of the sum is

$$p_{w+n}(w + n) = p_w(w) * p_n(n) \tag{6.14}$$

where $*$ denotes convolution and $p_w(w)$ is computed using the pdf of a function of a random variable [Papoulis, 1991]

$$p_w(w) = \sum_{x_i, w=g(x_i)} \frac{p_x(x_i)}{|g'(x_i)|} \tag{6.15}$$

and $g'(x)$ denotes the derivative of $g(x)$. Once $p_{w+n}(w + n)$ is known, we can compute $h(g(x) + n)$ using (6.5) and finally $I(x; y)$ using (6.4). The expression for $I(x; y)$ is not closed form for a general $g(x)$, but can be obtained using numerical integration. The amount of distortion introduced by the nonlinearity depends only on the ratio A/σ_x, i.e. the *ClipLevel*, that was defined in (3.96).

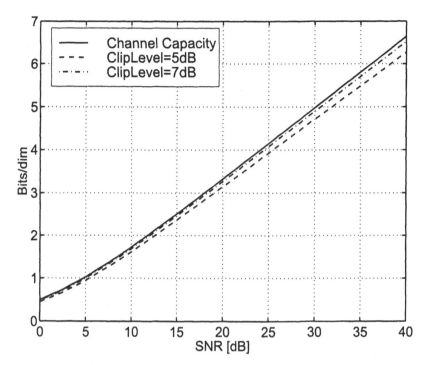

Figure 6.2: Channel capacity and mutual information for a *ClipLevel* of 5 *dB* and 7 *dB* for the Soft Limiter nonlinearity.

Figure 6.2 compares the mutual information $I(x; y)$ for a *ClipLevel* of 5 *dB* and 7 *dB* to the channel capacity C, as a function of SNR for the SL nonlinearity. Note that these mutual information rates can only be achieved in the context of a code with unconstrained complexity and decoding delay, and therefore should be considered as upper-bounds for the performance of any practical communication system. To better capture the small mutual information loss, the mutual information relative to the perfectly linear case is plotted in Figure 6.3, i.e. $I(x; y)/C$, for a *ClipLevel* of 5 *dB*, 7 *dB* and 9 *dB* for the SL nonlinearity. As seen from the figure, the reduction in mutual information due to the nonlinearity is small. Figure 6.4 compares the relative mutual information for the SSPA nonlinearity with the same *ClipLevels* and a smoothness factor

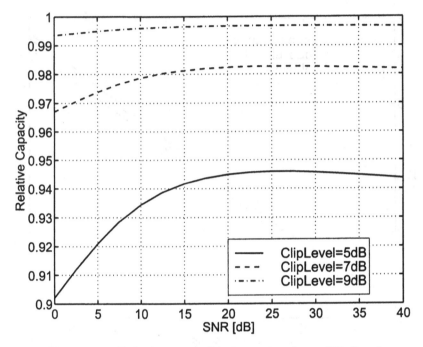

Figure 6.3: Relative mutual information for a *ClipLevel*
of 5 *dB*, 7 *dB* and 9 *dB* for the SL
nonlinearity.

as defined in (3.93) of $p = 2$. Overall, $I(x; y)$ is lower for the SSPA model, but it is still close to the channel capacity C for *ClipLevels* > 9 *dB* and moderate to high SNR.

2.   MAXIMUM LIKELIHOOD (ML) DETECTION

This section shows that the receiver can accurately estimate the transmitted vector from the distorted, filtered and noisy received vector by performing a simplified Maximum Likelihood (ML) detection. The following analysis assumes that the nonlinearity is at the transmitter. Nevertheless, the same analysis can be replicated, with minor changes, for the case where the distortion is caused by the channel or the receiver. First, the effect of the nonlinearity on the received DMT/OFDM vector is studied. Then, the degradation on maximum achievable rate that results from treating the distortion as an AWGN term is quantified.

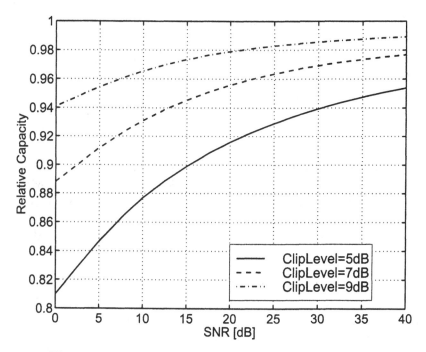

Figure 6.4: Relative mutual information for a *ClipLevel*
of 5 *dB*, 7 *dB* and 9 *dB* for the Solid-State
Power Amplifier nonlinearity.

Since the distortion is a deterministic function of the transmit vectors, treating it as an additive independent term is clearly inaccurate. Next, the ML receiver is formulated. Because of its exponential complexity, an iterative, approximate ML solution is proposed. The mathematical derivation will use a discrete-time model of the transceiver. If the only nonlinearity is in the discrete-time domain (e.g. DAC), this analysis is exact. However, if the nonlinearity is present in the continuous-time domain, this analysis will only be an approximation, but it is accurate if the oversampling of the discrete-time model is increased to include the spectral regrowth from the nonlinearity. In practice, since most of the energy from the nonlinearity will be in the third-order inter-modulation distortion, 3× oversampling will often be sufficient.

The output of the memoryless nonlinearity can be written as

$$x_n^g = g(x_n) = k^g x_n + d_n^{(\mathbf{X}, g)} \tag{6.16}$$

Depending on the transmitter structure and the location of the nonlinearity, the signal x_n can be replaced by the any of the following signals $x[n/L]$, $z_F[n/L]$, $z_T[n/L]$ $x_D(t)$ or $x_C(t)$ defined in Chapter 3. The constant k^g is chosen to minimize the MSE term $E\{|g(x_n) - k^g x_n|^2\}$ [Friese, 1998]. Therefore, the sequence $d_n^{(\mathbf{X},g)} = g(x_n) - k^g x_n$ is the sequence that contains the minimum distortion energy and is also uncorrelated with x_n. The notation $d_n^{(\mathbf{X},g)}$ is used to stress that d_n is a function of the multicarrier QAM data vectors $\mathbf{X}$ and the nonlinear function $g(\cdot)$. If the nonlinearity is present after the digital filter for $z_F[n/L]$, or after the analog filter for $x_C(t)$, $d_n^{(\mathbf{X},g)}$ will also depend on $p[n/L]$ and/or $p(t)$. It can be shown [Friese, 1998] that for the SL and SSPA nonlinearities, $k^g \to 1$ for $ClipLevels > 7\ dB$. Thus (6.16) can be approximated by

$$x_n^g = g(x_n) = x_n + d_n^{(\mathbf{X},g)} \tag{6.17}$$

In general, each distortion sample in the sequence $d_n^{(\mathbf{X},g)}$ can influence several received multicarrier vectors, but when the channel ISI is shorter than the CP, the distortion can be limited in its effect to a single multicarrier symbol. For example, if the nonlinearity is performed on the sequence $x[n/L]$ (see (3.11)), since the distortion is a deterministic function of $x[n/L]$, then $d_n^{(\mathbf{X},g)}$ and $x^g[n/L]$ will also satisfy the cyclic-prefix property. Therefore, the equivalent transmit QAM vector for the m-th symbol $\mathbf{X}_L^m$, including the nonlinear effect can be computed from the DFT of (6.17) for the case $x_n^g = x^g[n/L]$ over the discrete-time interval $(m(N+\nu)L, NL - 1 + m(N+\nu)L))$, i.e.

$$DFT(x_n^{m,g}) = \mathbf{X}_L^{m,g} = \mathbf{X}_L^m + \mathbf{D}_L^{(\mathbf{X}^m,g)} = \tag{6.18}$$

$$= \begin{cases} X_k^m + D_{L,k}^{(\mathbf{X}^m,g)}, & k = 0, \dots, \frac{N}{2} - 1 \\ X_{k-N(L-1)}^m + D_{L,k}^{(\mathbf{X}^m,g)}, & k = NL - \frac{N}{2}, \dots, NL - 1 \\ D_{L,k}^{(\mathbf{X}^m,g)}, & k = \frac{N}{2}, \dots, NL - \frac{N}{2} - 1 \end{cases} \tag{6.19}$$

The distortion terms $D_{L,k}^{(\mathbf{X}^m,g)}$, $k = \frac{N}{2}, \dots, NL - \frac{N}{2} - 1$ represent the out-of-band distortion and can be minimized with clip windowing [Pauli and Kuchenbecker, 1997, van Nee and de Wild, 1998, Pauli and Kuchenbecker, 1998] or by using smoother nonlinearities. If the nonlinearity is present on the Nyquist rate samples (e.g. a saturating DAC with $L = 1$), then, these terms will not be present. For the case of nonlinear distortion on $x_C(t)$ or $z_T[n/L]$, (6.19) will also apply since the cyclic prefix structure is preserved, but will only be an approximation when digital $(p[n/L])$ or analog $(p(t))$ filtering precedes the nonlinearity. The decomposition of the sequential signal transmission into independent symbols, allows working with each multicarrier vector independently, and the symbol index m may be dropped to simplify the notation. So far, this section has described the equivalent nonlinear transmit symbols and will now proceed to describe the receiver structures. We will assume that the out-of-band terms are filtered at the receiver input and are no longer available for decoding. This is realistic since these adjacent bands will typically be used by other transceivers. To further simplify notation, this section will also drop the oversampling index L from all derivations but it must be included in all the equations since the equivalent distortion vector $D_L^{(\mathbf{X}^m, g)}$ depends on the oversampling. Calling H_k the DFT of the channel impulse response $h[n]$, the received vector is [Bingham, 1990]

$$Y_k^g = H_k(X_k + D_k^{(\mathbf{X}, g)}) + N_k \qquad (6.20)$$

where N_k is the additive white Gaussian noise component for tone k. With this noise model, the ML receiver must solve the following equation

$$\hat{\mathbf{X}} = \arg\min_{\forall \bar{X}} \sum_{k=0}^{N-1} (H_k(\bar{X}_k + D_k^{(\bar{\mathbf{X}}, g)}) - Y_k^g)^2 \qquad (6.21)$$

Equation (6.21) implicitly assumed that all coding is performed within a single DMT/OFDM symbol. For example, if the outputs of a trellis coded modulator were interleaved and mapped into several DMT/OFDM symbols, (6.21) should include all the DMT/OFDM symbols in the codeword. To simplify the derivations, we will not consider multiple DMT/OFDM symbol codes.

Using vector notation, (6.21) can be written more compactly as

$$\hat{\mathbf{X}} = \arg\min_{\bar{\mathbf{X}}} \|\mathbf{H} \circ (\bar{\mathbf{X}} + \mathbf{D}^{(\bar{\mathbf{X}}, g)}) - \mathbf{Y}^g\|^2 \qquad (6.22)$$

where the vector operation $\mathbf{u} = \mathbf{v} \circ \mathbf{w}$ denotes the element by element vector product ($\mathbf{u} = [v_1 w_1, v_2 w_2, \cdots, v_N w_N]^T$). If we substitute $\mathbf{Y}^g$ from (6.20) in (6.22), we have

$$\hat{\mathbf{X}} = \arg\min_{\bar{\mathbf{X}}} \|\mathbf{H} \circ (\bar{\mathbf{X}} + \mathbf{D}^{(\bar{\mathbf{X}},g)}) - \mathbf{H} \circ (\mathbf{X} + \mathbf{D}^{(\mathbf{X},g)}) - \mathbf{N}\|^2 \qquad (6.23)$$

In general, the terms $\mathbf{D}^{(\mathbf{X},g)}$ and $\mathbf{D}^{(\bar{\mathbf{X}},g)}$ are complicated nonlinear functions of $\mathbf{X}$ and $\bar{\mathbf{X}}$. Therefore, solving for (6.23) exactly requires computing $\mathbf{D}^{(\bar{\mathbf{X}},g)}$ for all possible transmit symbol vectors $\bar{\mathbf{X}}$, which leads to an algorithm with exponential complexity. To avoid this huge complexity, practical receivers will not compute the term $\mathbf{D}^{(\bar{\mathbf{X}},g)}$ and instead assume the distortion term $\mathbf{D}^{(\mathbf{X},g)}$, which is uncorrelated with $\mathbf{X}$, can be approximated as an AWGN term. In this case, the ISI channel reduces to N independent subchannels:

$$Y_k^g = H_k X_k + \hat{N}_k \qquad (6.24)$$

where

$$\hat{N}_k = H_k D_k^{(\mathbf{X},g)} + N_k \qquad (6.25)$$

For this case, the receiver is approximated by a standard DMT/OFDM receiver that has an additional noise term. For large DMT/OFDM symbol sizes, and low to moderate values of *ClipLevel*, this additional noise term is approximately Gaussian [Friese, 1998]. Under these assumptions, the achievable data rate per dimension (real or imaginary part) on subchannel k is:

$$R_k \approx \frac{1}{2} \log_2 \left(1 + \frac{\sigma_{X,k}^2}{\sigma_{D,k}^2 + \sigma_{N,k}^2} \right) \qquad (6.26)$$

where the variable $\sigma_{X,k}^2 = |H_k|^2 E\{|X_k|^2\}$ is the received signal power, $\sigma_{D,k}^2 = |H_k|^2 E\{|D_k|^2\}$ is the received distortion power and $\sigma_{N,k}^2 = E\{|N_k|^2\}$ is the received noise in the k-th subchannel. Plots in Figure 6.5 are for the SL nonlinearity at *ClipLevel*s of 5 *dB*, 7 *dB* and 9 *dB*. These plots show the effective data rates given by (6.26) when the distortion is present at the Nyquist rate samples. These achievable rates are significantly lower than the (optimal) mutual information values shown in Figure 6.2, especially at moderate to high SNR. This motivates us to design a quasi-ML receiver that can achieve performance closer to the optimal values.

If the receiver can compute the transmit distortion $\mathbf{D}^{(\mathbf{X},g)}$, the exact ML problem will be simplified, since this term $\mathbf{D}^{(\mathbf{X},g)}$ can be canceled

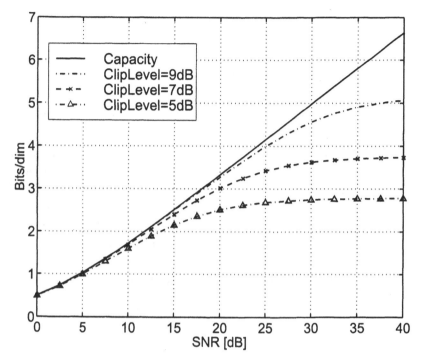

Figure 6.5: Channel capacity and practical data rates for
a *ClipLevel* of 5 *dB*, 7 *dB* and 9 *dB* with a
SL if the distortion is assumed to be AWGN.

in (6.23), and the deterministic term $\mathbf{D}^{(\bar{\mathbf{X}},g)}$ will be unnecessary. That
is, (6.23) reduces to

$$\hat{\mathbf{X}} = \arg\min_{\bar{\mathbf{X}}} \|\mathbf{H} \circ \bar{\mathbf{X}} - \mathbf{H} \circ (\mathbf{X} + \mathbf{D}^{(\mathbf{X},g)} - \mathbf{D}^{(\mathbf{X},g)}) - \mathbf{N}\|^2 \qquad (6.27)$$

which simplifies to:

$$\hat{\mathbf{X}} = \arg\min_{\bar{\mathbf{X}}} \|\mathbf{H} \circ \bar{\mathbf{X}} - \mathbf{H} \circ \mathbf{X} - \mathbf{N}\|^2 \qquad (6.28)$$

In other words, the ML decoder reduces to the standard linear case, in
terms of complexity and performance.

All the following assumes uncoded transmission, but the proposed
simplified algorithms can be applied to the coded case with minor modifications. For the uncoded case, this vector optimization problem can

be decomposed into N, ML scalar problems:

$$\hat{X}_k = \arg\min_{\bar{X}_k}(H_k\bar{X}_k - H_kX_k - N_k)^2 \qquad (6.29)$$

$$= \arg\min_{\bar{X}_k}\left(H_k\left[\bar{X}_k - \frac{Y_k^g}{H_k} + D_k^{(\mathbf{X},g)}\right]\right)^2 \qquad (6.30)$$

Therefore, if $D_k^{(\mathbf{X},g)}$ is known, the ML receiver simply chooses the value $\hat{X}_k$ that is closest to $Y_k^g/H_k - D_k^{(\mathbf{X},g)}$.

Thus, if the receiver has access to $\mathbf{D}^{(\mathbf{X},g)}$, the system will not exhibit any degradation with respect to the undistorted system, and the additional decoding complexity is minimal. One solution to obtain $\mathbf{D}^{(\mathbf{X},g)}$ will be to transmit side information that will enable us to reconstruct $\mathbf{D}^{(\mathbf{X},g)}$. In general, an accurate representation of $\mathbf{D}^{(\mathbf{X},g)}$ may require large amounts of information which can in turn reduce the throughput of our communications link considerably. However, a simple parametric model may convey enough information to provide a good replica of $\mathbf{D}^{(\mathbf{X},g)}$.

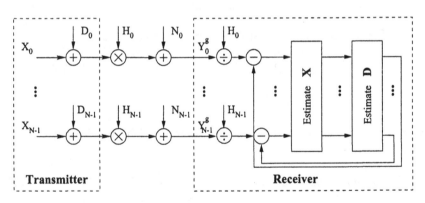

Figure 6.6: Iterative (quasi-ML) nonlinear distortion canceler.

Alternatively, the receiver can compute an estimate of $\mathbf{D}^{(\mathbf{X},g)}$, denoted $\hat{\mathbf{D}}^{(\mathbf{X},g)}$, from the received vector $\mathbf{Y}^g$ as shown in Figure 6.6 if the receiver knows the transmit nonlinear function $g(\cdot)$ and has an estimate of the transmit QAM vector $\mathbf{X}$. The accuracy for the estimate of $\mathbf{D}^{(\mathbf{X},g)}$ will depend on the estimate of $\mathbf{X}$, but in general the receiver performance will improve if we reduce the energy of the distortion vector,

$$E\{|\mathbf{D}^{(\mathbf{X},g)} - \hat{\mathbf{D}}^{(\mathbf{X},g)}|^2\} < E\{|\mathbf{D}^{(\mathbf{X},g)}|^2\} \qquad (6.31)$$

This property is demonstrated from the numerical examples in Section 3. for the simple uncoded case. If $\mathbf{D}^{(\mathbf{X},g)} \neq \hat{\mathbf{D}}^{(\mathbf{X},g)}$, we will not achieve perfect distortion cancellation in (6.27). However, if (6.31) is satisfied, the distortion will be smaller and the receiver will get a second improved estimate of $\mathbf{X}$ that it can use to reestimate $\mathbf{D}^{(\mathbf{X},g)}$.

The following describes a simple iterative algorithm based on hard decoding of the received vector but other alternatives are possible. For example, if the transceiver includes some error correction, the estimate of $\mathbf{X}$ will be more accurate and so will $\hat{\mathbf{D}}^{(\mathbf{X},g)}$. Performing hard decoding on the received vector $\mathbf{Y}^g$ will result in a simple estimate of the transmitted symbol $\mathbf{X}$:

$$\mathbf{X}^{(1)} = \left[\langle \frac{Y_0^g}{H_0} \rangle \cdots \langle \frac{Y_{N-1}^g}{H_{N-1}} \rangle \right]^T = \langle \mathbf{Y}^g / \mathbf{H} \rangle \qquad (6.32)$$

where $\langle A_k \rangle$ denotes hard decoding, i.e. the operation of selecting the constellation point of tone k that is closest to A_k. From this estimate, we can compute, $\mathbf{x}^{(1)} = IDFT(\mathbf{X}^{(1)})$, and thus the estimate for $\mathbf{D}^{(\mathbf{X},g)}$ is:

$$\mathbf{D}^{(\mathbf{X}^{(1)},g)} = DFT(d_n^{(\mathbf{X}^{(1)},g)}) = DFT(g(x_n^{(1)}) - x_n^{(1)}) \qquad (6.33)$$

If the SER is $p \ll 1$, a large fraction of the hard-decoded values will be correct since $Prob\{X_k^{(1)} = X_k\} = 1-p$. Therefore, $\mathbf{D}^{(\mathbf{X},g)} \approx \mathbf{D}^{(\mathbf{X}^{(1)},g)}$. Although the estimates may degrade from errors, the receiver can use these estimates to reduce the nonlinear distortion term in (6.23) in an iterative fashion. In this case, (6.27) becomes:

$$\mathbf{X}^{(q+1)} = \arg\min_{\bar{\mathbf{X}}} \|\mathbf{H} \circ \bar{\mathbf{X}} - \mathbf{H} \circ (\mathbf{X} + \mathbf{D}^{(\mathbf{X},g)} - \mathbf{D}^{(\mathbf{X}^{(q)},g)}) - \mathbf{N}\|^2 \quad (6.34)$$

or equivalently:

$$\mathbf{X}^{(q+1)} = \arg\min_{\bar{\mathbf{X}}} \|\bar{\mathbf{X}} - (\frac{\mathbf{Y}^g}{\mathbf{H}} - \mathbf{D}^{(\mathbf{X}^{(q)},g)})\|^2 \qquad (6.35)$$

For the uncoded case, the proposed iterative quasi-ML algorithm can be summarized in three steps, starting with $q = 1$ and $\mathbf{D}^{(\mathbf{X}^{(0)},g)} = \mathbf{0}$.

$$\mathbf{X}^{(q)} = \langle \frac{\mathbf{Y}^g}{\mathbf{H}} - \mathbf{D}^{(\mathbf{X}^{(q-1)},g)} \rangle, \tag{6.36}$$

$$x_n^{(q)} = \mathbf{x}^{(q)} = IDFT(\mathbf{X}^{(q)}) \tag{6.37}$$

$$\mathbf{D}^{(\mathbf{X}^{(q)},g)} = DFT(d_n^{(\mathbf{X}^{(q)},g)}) = DFT(g(x_n^{(q)}) - x_n^{(q)}), \tag{6.38}$$

The first step estimates the data vector $\mathbf{X}^{(q)}$ from the received vector and the previous distortion estimate $\mathbf{D}^{(\mathbf{X}^{(q-1)},g)}$, and the last two steps update the distortion estimate from the new data vector estimate.

3.   NUMERICAL RESULTS

Figure 6.7 and Figure 6.8 show the performance of the proposed iterative-ML algorithm for a DMT transceiver in an ADSL application. The transmitted multicarrier signal has 256 independent tones, i.e. $N = 512$ and the channel is a typical ADSL channel with exponentially decreasing gains, such that the number of bits per channel varies from a maximum of 10 bits/tone down to 2 bits/tone. The results in Figure 6.7 are for the SL nonlinearity case and with *ClipLevel = 9 dB*. Therefore, the PAR at the transmitter is about 9 *dB*. The top-most curve, which is denoted "Original" in the plot, is the SER for an uncoded system without clip estimation. The dashed curve under it is the estimated SER if the nonlinear distortion is assumed a Gaussian approximation. The three immediate curves below it (lower SER) correspond to the SER after the first, second and third iteration of the proposed iterative quasi-ML algorithm. Finally, the bottom-most curve is the SER for an undistorted transmitter, that is, a *perfectly linear* transmitter. From the plot, the difference between the ideal linear SER curve and the SER curve after the third iteration of the proposed algorithm is negligible. At a normalized[2] SNR of 14 *dB*, the SER reduction is more than 4 orders of magnitude.

The results in Figure 6.8 are for the SSPA nonlinearity case with a smoothness factor, $p = 2$ and *ClipLevel = 11 dB*. Therefore, the PAR

---

[2]The subchannel/tone SNR_k have been normalized so that they can all be shown in the same plot. The x-axis is actually received $d_{min,k}/2\sigma_{N,k}$

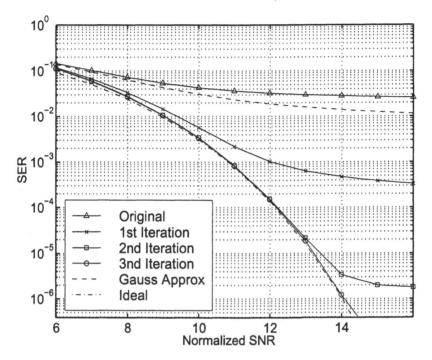

Figure 6.7: Performance of the iterative-ML algorithm
for a SL nonlinearity when $N = 512$, $L = 1$
and $ClipLevel = 9\ dB$.

at the transmitter is about 11 dB, but for this specific nonlinearity, the
transmitted signal is not linear over the range $(-A, A)$. The notation
is the same as in Figure 6.7, but only 2 iterations are necessary to be
within .5 dB of the ideal case.

In Figure 6.9 and Figure 6.10, we increase the number of independent
tones to 2048. This is the number of tones proposed for the VDSL-DMT
standard. As in the previous two figures, the channel has exponentially decreasing gains, such that the number of bits per channel varies
from a maximum of 10 bits/tone down to 2 bits/tone. In both Figure 6.9 and Figure 6.10, the nonlinear operation was performed on the
2 times oversampled signal $z_T[n/2]$. In Figure 6.9, the Gaussian clip
windowing algorithm described in [Pauli and Kuchenbecker, 1997, van
Nee and de Wild, 1998, Pauli and Kuchenbecker, 1998] was applied with
$ClipLevel = 9\ dB$. The iterative-ML receiver converges to the ideal case
by the second iteration. The results in Figure 6.10, which are for the

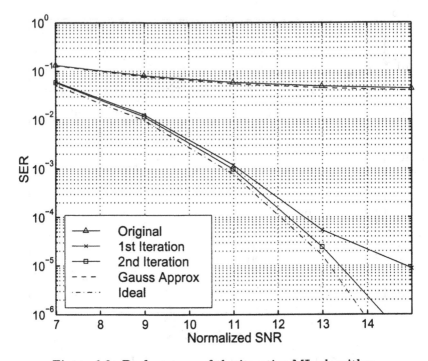

Figure 6.8: Performance of the iterative-ML algorithm
for a SSPA nonlinearity when $N = 512$,
$L = 1$ and $ClipLevel = 11 \ dB$.

SL nonlinearity case and $ClipLevel = 8 \ dB$, also indicate that the ideal case is approached after two iterations. Comparing Figure 6.7 and Figure 6.10, we observe that increasing the DFT size improves the performance, since the latter plots converge faster despite having a lower $ClipLevel$ and therefore more distortion.

4.   CONCLUSIONS

Since multicarrier symbols exhibit approximately Gaussian distributed, time-domain waveforms with relatively high PAR, in many instances, the transmitter, the channel or the receiver are operating in a nonlinear region to avoid extremely large transmit power back-offs that would degrade the received SNR. The optimal Maximum Likelihood (ML) receiver was derived, but unfortunately, due to the nonlinear distortion, the ML decoder often has an exponential complexity. To avoid this

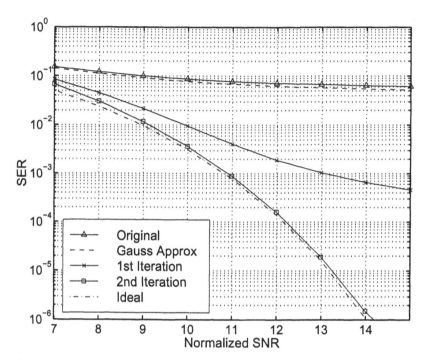

Figure 6.9: Performance of the iterative-ML algorithm
for the Gaussian clip windowing nonlinearity
with Gaussian clip windowing when
$N = 4096$, $L = 2$ and $ClipLevel = 9$ dB.

extremely large complexity, we described a simple algorithm that iteratively estimates the nonlinear distortion and reduces the exponential-ML to the standard ML without nonlinear distortion. This structure was able to reduce the SER by several orders of magnitude in a typical ADSL channel at the expense of two additional FFT of complexity. These additional FFT have a small number of nonzero values and can be implemented efficiently.

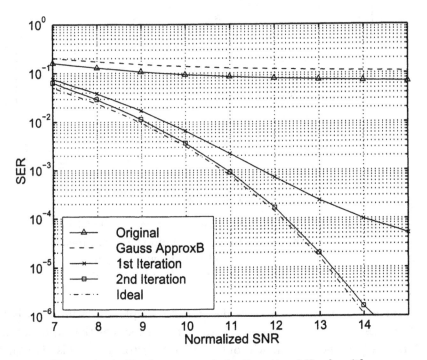

Figure 6.10: Performance of the iterative-ML algorithm
for a SL nonlinearity when $N = 4096$,
$L = 2$ and *ClipLevel* $= 8$ dB.

Chapter 7

SUMMARY AND CONCLUSIONS

1.   BOOK SUMMARY

THIS BOOK FORMULATED THE PAR PROBLEM for multicarrier modulation and proposed three new methods for PAR reduction. Chapter 2 introduced the basics of multicarrier modulation, where the focus was on the discrete-time model. Although Vector Coding was described, most of the formulation was based on the two most commonly used multicarrier modulation in practice, namely, Discrete MultiTone (DMT) and Orthogonal Frequency Division Multiplexing (OFDM).

Chapter 3 described real and complex multicarrier signals using continuous-time and discrete-time formulations, and formally introduced the concept of PAR. This chapter also derived novel bounds on the PAR of the continuous-time symbols given the discrete-time samples. Since the PAR reduction methods described in Chapters 4 and 5 are simplified when operating on discrete-time symbols, these bounds can be used to predict the continuous-time PAR. Furthermore, several common nonlinear models were presented and their effects on multicarrier demodulation were evaluated. The main degradation effects from processing multicarrier signals with nonlinear distortion are spectral re-growth and BER increase. A new expression for evaluating BER for ideal soft limiters was also derived.

Chapters 4, 5 and 6 introduced three new structures for PAR reduction. The first two prevent distortion by reducing the PAR on the digital signal prior to any nonlinear device such as a DAC or a power amplifiers. The last structure corrects for transmitter nonlinear distortion at the receiver when the nonlinear function is known.

Chapter 4 presented the first new distortionless PAR reduction structure denoted as Tone Reservation. The exact solution to this PAR minimization problem, as well as efficient suboptimal solutions were described, followed by a number of simulation results for practical multicarrier systems. Since this is a prevention method, most of the additional complexity introduced is at the transmitter. The simple structure introduced in the PAR reduction vectors does not add any complexity at the receiving end of the transmission link.

Chapter 5 described a second distortionless PAR reduction structure denoted as Tone Injection. Since the exact solution to this PAR minimization problem has non-polynomial complexity, limits on maximum PAR reduction were derived. Efficient suboptimal solutions that achieve near optimal performance were proposed. Similar to the Tone Reservation method, most of the complexity for reducing the PAR is introduced at the transmitter. The additional complexity at the receiver is a simple modulo operation of the demodulated complex vectors.

Chapter 6 described a structure that reduces the PAR by applying a saturating nonlinearity at the transmitter and correcting for nonlinear distortion at the receiver. The performance of multicarrier transmission in the presence of nonlinear distortion was evaluated and theoretical Mutual Information limits were derived. The optimal maximum likelihood receiver and an efficient demodulator based on the maximum likelihood receiver were also described.

References

[Andreoli et al., 1997] Andreoli, S., McClure, H. G., Banelli, P., and Cacopardi, S. (1997). Digital linearizer for RF amplifiers. *IEEE Trans Broadcasting*, 43(1):12–19.

[ANSI, 1995] ANSI (1995). *Network and Customer Installation Interfaces-Asymmetric Digital Subscriber Line (ADSL) Metallic Interface*. T1.413.

[Aslanis, 1989] Aslanis, J. T. (1989). *Coding for Communications Channels with Memory*. Ph.D. Thesis, Stanford University.

[Bäuml et al., 1996] Bäuml, R. W., Fischer, R. F. H., and Huber, J. B. (1996). Reducing the peak-to-average power ratio of multicarrier modulation by selected mapping. *Electronics Letters*, 32(22):2056–2057.

[Bingham, 1990] Bingham, J. (1990). Multicarrier modulation for data transmission: an idea whose time has come. *IEEE Commun Mag*.

[Boyd, 1986] Boyd, S. (1986). Multitone signals with low crest factor. *IEEE Trans Circuit Syst*, CAS-33(10):1018–1022.

[Boyd and Vandenberghe, 1997] Boyd, S. and Vandenberghe, L. (1997). Lecture notes for introduction to convex optimization with engineering applications. Electrical Engineering Department, Stanford University, CA, 1997.

[Boyd et al., 1994] Boyd, S., Vandenberghe, L., and Grant, M. (1994). Efficient convex optimization for engineering design. In *Proceedings IFAC Symposium on Robust Control Design*, Rio de Janeiro, Brazil.

[Campello de Sousa, 1998] Campello de Sousa, J. (1998). Optimal discrete bit loading for multicarrier modulation systems. In *Proc IEEE Intl Symp Inform Theory*, page 193, Boston, MA.

[Campello de Sousa, 1999] Campello de Sousa, J. (1999). *Discrete Bit Loading for Multicarrier Modulation Systems*. Ph.D. Thesis, Stanford University.

[Chow et al., 1997a] Chow, J. S., Bingham, J. A. C., and Flowers, M. S. (1997a). Mitigating clipping noise in multicarrier systems. In *Proc IEEE Intl Conf Commun*, pages 715–719, Montreal, Canada.

[Chow et al., 1997b] Chow, J. S., Bingham, J. A. C., Flowers, M. S., and Cioffi, J. M. (1997b). Mitigating clipping and quantization effects in digital transmission systems. U.S. Patent No. 5,623,513 (Apr. 22, 1997).

[Chow et al., 1991a] Chow, J. S., Tu, J. C., and Cioffi, J. M. (1991a). A discrete multitone transceiver system for HDSL applications. *IEEE J Sel Areas Commun*, 9(6):895–908.

[Chow, 1993] Chow, P. S. (1993). *Bandwidth Optimized Digital Transmission Techniques for Spectrally Shaped Channels with Impulse Noise*. Ph.D. Thesis, Stanford University.

[Chow et al., 1991b] Chow, P. S., Tu, J. C., and Cioffi, J. M. (1991b). Performance evaluation of a multichannel transceiver system for ADSL and VHDSL services. *IEEE J Sel Areas Commun*, 9(6):909–919.

[Cimini, Jr., 1985] Cimini, Jr., L. J. (1985). Analysis and simulation of a digital mobile channel using orthogonal frequency division multiplexing. *IEEE Trans Commun*, COM-33(7):665–75.

[Cioffi, 1991] Cioffi, J. M. (1991). A multicarrier primer. *ANSI Document, T1E1.4 Technical Subcommittee*, (91-157).

[Cioffi, 2000a] Cioffi, J. M. (2000a). EE 379A course notes. Stanford University, Stanford, CA, http://www.stanford.edu/class/ee379a/reader.html.

[Cioffi, 2000b] Cioffi, J. M. (2000b). EE 379C advanced digital communications course notes. Stanford University, Stanford, CA, http://www.stanford.edu/class/ee379c/reader.html.

[Cioffi et al., 1995] Cioffi, J. M., Dudevoir, G., Eyuboglu, V., and Forney, G. D. (1995). MMSE decision-feedback equalizers and coding - Part I: Equalization results. *IEEE Trans Commun*, 43(10):2582–94.

[Cover and Thomas, 1991] Cover, T. M. and Thomas, J. A. (1991). *Elements of Information Theory*. John Wiley & Sons.

[D'Andrea et al., 1996] D'Andrea, A. N., Lottici, V., and Reggiannini, R. (1996). RF power amplifier linearization through amplitude and phase predistortion. *IEEE Trans Commun*, COM-44(11):1477–1484.

[Davis and Jedwab, 1997] Davis, J. A. and Jedwab, J. (1997). Peak-to-mean power control and error correction for OFDM transmission using Golay sequences and Reed-Muller codes. *Electronics Letters*, 33(4):267–268.

[Djokovic, 1997] Djokovic, I. (1997). PAR reduction without noise enhancement. *ANSI Document, T1E1.4 Technical Subcommittee*, (97-270):1–3.

[Dudevoir, 1989] Dudevoir, G. P. (1989). *Equalization Techniques for High Rate Digital Transmission on Spectrally Shaped Channels.* Ph.D. Thesis, Stanford University.

[EN300744, 1997] EN300744 (1997). *Digital Video Broadcasting (DVB); Framing structure, channel coding and modulation for digital terrestrial television.*

[ETSI, 1995] ETSI (1995). *Radio broadcasting systems; Digital Audio Broadcasting (DAB) to mobile, portable and fixed receivers.* Valbonne, France.

[Friese, 1996] Friese, M. (1996). Multicarrier modulation with low peak-to-average power ratio. *Electronics Letters*, 32(8):713–714.

[Friese, 1997a] Friese, M. (1997a). OFDM signals with low crest-factor. In *Proc IEEE GlobeCom*, volume 1, pages 290–294, Phoenix, AZ.

[Friese, 1997b] Friese, M. (1997b). Multitone signals with low crest factor. *IEEE Trans Commun*, 45(10):1338–1344.

[Friese, 1998] Friese, M. (1998). On the degradation of OFDM-signals due to peak-clipping in optimally predistorted power amplifiers. In *Proc IEEE GlobeCom*, volume 2, pages 939–944, Sydney, Australia.

[Gallager, 1968] Gallager, R. G. (1968). *Information Theory and Reliable Communication.* Wiley, New York.

[Gatherer and Polley, 1997] Gatherer, A. and Polley, M. (1997). Controlling clipping probability in DMT transmission. In *Proc. of 31st Asilomar Conf. on Signals, Systems, and Computers*, Pacific Grove, CA.

[Golay, 1961] Golay, M. J. E. (1961). Complementary series. *IRE Trans Inform Theory*, IT-7(7):82–87.

[Greenstein and Fitzgerald, 1981] Greenstein, L. J. and Fitzgerald, P. J. (1981). Phasing multitone signals to minimize peak factors. *IEEE Trans Commun*, COM-29(7):1072–1074.

[Gross and Veeneman, 1994] Gross, R. and Veeneman, D. (1994). SNR and spectral properties for a clipped DMT ADSL signal. In *Proc IEEE Intl Conf Commun*.

[Henkel and Wagner, 1997] Henkel, W. and Wagner, B. (1997). Trellis shaping for reducing the peak-to-average ratio of multitone signals. In *Proc IEEE Intl Symp Inform Theory*, page 516, Ulm, Germany.

[Hoo et al., 1998a] Hoo, L. M. C., Tellado, J., and Cioffi, J. M. (1998a). Dual QoS loading algorithms for dmt systems offering cbr and vbr services. In *Proc IEEE GlobeCom*, pages 25–30, Sydney, Australia.

[Hoo et al., 1998b] Hoo, L. M. C., Tellado, J., and Cioffi, J. M. (1998b). Dual QoS loading algorithms for multicarrier systems offering different cbr services. In *Proc IEEE PIMRC*, pages 278–282, Boston, MA.

[Hoo et al., 1999] Hoo, L. M. C., Tellado, J., and Cioffi, J. M. (1999). Discrete dual QoS loading algorithms for multicarrier systems. In *Proc IEEE Intl Conf Commun*, Vancouver, Canada.

[Hughes-Hartogs, 1989] Hughes-Hartogs, D. (1989). Ensemble modem structure for imperfect transmission media. U.S. Patents Nos. 4,679,227 (July 1987), 4,731,816 (March 1988) and 4,833,796 (May 1989).

[ITU, 1999] ITU (1999). *Draft Recomendation G.992*. http://ties.itu.int/u/tsg15/sg15/wp1/q4/.

[Jeon et al., 1997] Jeon, W. G., Chang, K. H., and Cho, Y. S. (1997). An adaptive data predistorter for compensation of nonlinear distortion in OFDM systems. *IEEE Trans Commun*, COM-45(10):1167–1171.

[Jones and Wilkinson, 1996] Jones, A. E. and Wilkinson, T. A. (1996). Combined coding for error control and increased robustness to system nonlinearities in OFDM. In *Proc IEEE Vehicular Tech Conf*, volume 2, pages 904–908, Atlanta, GA.

[Jones et al., 1994] Jones, E., Wilkinson, T. A., and Barton, S. K. (1994). Block coding scheme for reduction of peak to mean envelope power ratio of multicarrier transmission schemes. *Electronics Letters*, 30(25):2098–2099.

[Kasturia, 1989] Kasturia, S. (1989). *Vector Coding for Digital Communication on Spectrally Shaped Channels*. Ph.D. Thesis, Stanford University.

[Kasturia et al., 1990] Kasturia, S., Aslanis, J., and Cioffi, J. M. (1990). Vector coding for partial-response channels. *IEEE Trans Inform Theory*, 36(4):741–762.

[Kschischang et al., 1998a] Kschischang, F. R., Narula, A., and Eyuboglu, V. (Apr 14-15, 1998a). A new approach to PAR control in DMT systems. *UAWG Contribution*, (TG/98-127):1–5.

[Kschischang et al., 1998b] Kschischang, F. R., Narula, A., and Eyuboglu, V. (May 11-14, 1998b). A new approach to PAR control in DMT systems. *ITU, Q4/15*, (NF-083):1–5.

[Li and Cimini, Jr., 1997] Li, X. and Cimini, Jr., L. J. (1997). Effects of clipping and filtering on the performance of OFDM. In *Proc IEEE Vehicular Tech Conf*, volume 3, pages 1634–1638, Phoenix, AZ.

[Mestdagh et al., 1994] Mestdagh, D. J. G., Spruyt, P., and Biran, B. (1994). Analysis of clipping effect in DMT-based ADSL systems. In *Proc IEEE Intl Conf Commun*.

[Mestdagh and Spruyt, 1996] Mestdagh, D. J. G. and Spruyt, P. M. P. (1996). A method to reduce the probability of clipping in DMT-based transceivers. *IEEE Trans Commun*, COM-44(10):1234–1238.

[Moose, 1994] Moose, P. (1994). A technique for orthogonal frequency division multiplexing frequency offset correction. *IEEE Trans Commun*, 42(10):2908–2914.

[Müller et al., 1997] Müller, S. H., Bäuml, R. W., Fischer, R. F. H., and Huber, J. B. (1997). OFDM with reduced peak-to-average power ratio by multiple signal representation. *Ann. Telecommun*, 52(1-2):58–67.

[Müller and Huber, 1997a] Müller, S. H. and Huber, J. B. (1997a). A comparison of peak power reduction schemes for OFDM. In *Proc IEEE GlobeCom*, volume 1, pages 1–5, Phoenix, AZ.

[Müller and Huber, 1997b] Müller, S. H. and Huber, J. B. (1997b). OFDM with reduced peak-to-average power ratio by optimum combination of partial transmit sequences. *Electronics Letters*, 33(5):368–369.

[Müller and Huber, 1997c] Müller, S. H. and Huber, J. B. (1997c). A novel peak power reduction scheme for OFDM. In *Proc IEEE PIMRC*, pages 1090–1094, Helsinki, Finland.

[Narahashi et al., 1995] Narahashi, S., Kumagai, K., and Nojima, T. (1995). Minimising peak-to-average power ratio of multitone signals using steepest descent method. *Electronics Letters*, 31(18):1552–1553.

[Narahashi and Nojima, 1994] Narahashi, S. and Nojima, T. (1994). New phasing scheme of n-multiple carriers for reducing peak-to-average power ratio. *Electronics Letters*, 30(17):1382–1382.

[O'Neill and Lopes, 1994] O'Neill, R. and Lopes, L. B. (1994). Performance of amplitude limited multitone signals. In *Proc IEEE Vehicular Tech Conf*, volume 3, pages 1675–1679, Stockholm, Sweden.

[O'Neill and Lopes, 1995] O'Neill, R. and Lopes, L. B. (1995). Envelope variations and spectral splatter in clipped multicarrier signals. In *Proc IEEE PIMRC*, volume 1, pages 71–75, Toronto, Canada.

[Oppenheim and Schafer, 1989] Oppenheim, A. V. and Schafer, R. W. (1989). *Discrete-Time Signal Processing*. Prentice-Hall.

[Papoulis, 1991] Papoulis, A. (1991). *Probability, Random Variables, and Stochastic Processes*. McGraw-Hill.

[Park and Powers, 1998] Park, I.-S. and Powers, E. J. (1998). Compensation of nonlinear distortion in OFDM systems using a new predistorter. In *Proc IEEE PIMRC*, pages 811–815, Boston, MA.

[Paterson, 1998] Paterson, K. G. (1998). Coding techniques for power controlled OFDM. In *Proc IEEE PIMRC*, pages 801–805, Boston, MA.

[Pauli and Kuchenbecker, 1997] Pauli, M. and Kuchenbecker, H. P. (1997). Minimization of the intermodulation distortion of a nonlinearly amplified OFDM signal. *Wireless Personal Commun*, 4(1):93–101.

[Pauli and Kuchenbecker, 1998] Pauli, M. and Kuchenbecker, H. P. (1998). On the reduction of the out-of-band radiation of OFDM-signals. In *Proc IEEE Intl Conf Commun*, pages 1304–1308, Atlanta, GA.

[Peled and Ruiz, 1980] Peled, A. and Ruiz, A. (1980). Frequency domain data transmission using reduced computational complexity algorithms. In *Proc IEEE Intl Conf Acoustics, Speech, and Sig Proc*, pages 964–967, Denver, CO.

[Popovic, 1991] Popovic, B. M. (1991). Synthesis of power efficient multitone signals with flat amplitude spectrum. *IEEE Trans Commun*, COM-39(7):1031–1033.

[Reimers, 1998] Reimers, U. (1998). Digital video broadcasting. *IEEE Commun Mag*, pages 104–110.

[Rowe, 1982] Rowe, H. E. (1982). Memoryless nonlinearities with Gaussian inputs: elementary results. *Bell Systems Tech J*, 61(7):1519–1525.

[Ruiz et al., 1992] Ruiz, A., Cioffi, J. M., and Kasturia, S. (1992). Discrete multiple tone modulation with coset coding for the spectrally shaped channel. *IEEE Trans Commun*, COM-40(6):1012–1029.

[Saleh, 1981] Saleh, A. A. M. (1981). Frequency-independent and frequency-dependent nonlinear models of TWT amplifiers. *IEEE Trans Commun*, COM-29(11):1715–1720.

[Santella and Mazzenga, 1998] Santella, G. and Mazzenga, F. (1998). A hybrid analytical-simulation procedure for performance evaluation in M-QAM-OFDM schemes in prescence of nonlinear distortions. *IEEE Trans Vehicular Tech*, 47(1):142–151.

[Schmidl and Cox, 1997] Schmidl, T. M. and Cox, D. C. (1997). Robust frequency and timing synchronization for OFDM. *IEEE Trans Commun*, 45(12):1613–1621.

[Shamai (Shitz) and Bar-David, 1995] Shamai (Shitz), S. and Bar-David, I. (1995). The capacity of average and peak-power-limited quadrature gaussian channels. *IEEE Trans Inform Theory*, 41(4):1060–1071.

[Shamai (Shitz) and Dembo, 1994] Shamai (Shitz), S. and Dembo, A. (1994). Bounds on the symmetric binary cutoff rate for dispersive gaussian channels. *IEEE Trans Commun*, COM-42(1):39–53.

[Shannon, 1948a] Shannon, C. E. (1948a). A mathematical theory of communication. *Bell Systems Tech J*, pages 379–423.

[Shannon, 1948b] Shannon, C. E. (1948b). A mathematical theory of communication. *Bell Systems Tech J*, pages 623–656.

[Shepherd et al., 1998] Shepherd, S., Orriss, J., and Barton, S. (1998). Asymptotic limits in peak envelope power reduction by redundant coding in orthogonal frequency-division multiplex modulation. *IEEE Trans Commun*, COM-46(1):5–10.

[Smith, 1969] Smith, J. G. (1969). *On the information capacity of peak and average power constrained Gaussian channels*. Ph.D. Thesis, Berkeley, CA.

[Tellado and Cioffi, 1997a] Tellado, J. and Cioffi, J. M. (1997a). PAR reduction for multicarrier transmission systems. U.S. Patent Pending (1997).

[Tellado and Cioffi, 1998a] Tellado, J. and Cioffi, J. M. (1998a). Efficient algorithms for reducing PAR in multicarrier systems. In *Proc IEEE Intl Symp Inform Theory*, page 191, Boston, MA.

[Tellado and Cioffi, 1998f] Tellado, J. and Cioffi, J. M. (1998f). Peak power reduction for multicarrier transmission. In *Proc IEEE Globe-Com Commun Theory MiniConf (CTMC)*, Sydney, Australia.

[Tellado and Cioffi, 1999b] Tellado, J. and Cioffi, J. M. (1999b). Maximum likelihood detection of nonlinearly distorted multicarrier symbols by iterative decoding. In *Proc IEEE GlobeCom Commun Theory MiniConf (CTMC)*, Rio de Janeiro, Brazil.

[Tellado and Cioffi, 2000a] Tellado, J. and Cioffi, J. M. (2000a). Maximum likelihood detection of nonlinearly distorted multicarrier symbols by iterative decoding. *To appear in IEEE Trans Commun*.

[Tellado and Cioffi, 2000b] Tellado, J. and Cioffi, J. M. (2000b). Tone reservation peak to average power reduction for multicarrier transmission. *Submitted to IEEE Trans Commun*.

[Tellado and Cioffi, 1998h] Tellado, J. and Cioffi, J. M. (Apr 20-24, 1998h). Revisiting DMT's PAR. *ETSI, TM6*, (TD08).

[Tellado and Cioffi, 1998b] Tellado, J. and Cioffi, J. M. (Aug, 1998b). Further results on peak-to-average ratio reduction. *ANSI Document, T1E1.4 Technical Subcommittee*, (98-252):1–8.

[Tellado and Cioffi, 1997b] Tellado, J. and Cioffi, J. M. (Dec 8, 1997b). PAR reduction in multicarrier transmission systems. *ANSI Document, T1E1.4 Technical Subcommittee*, (97-367):1–14.

[Tellado and Cioffi, 1998d] Tellado, J. and Cioffi, J. M. (Feb 9-20, 1998d). PAR reduction in multicarrier transmission systems. *ITU, Q4/15*, (D-150(WP 1/15)):1–14.

[Tellado and Cioffi, 1998e] Tellado, J. and Cioffi, J. M. (Jun, 1998e). PAR reduction with minimal or zero bandwidth loss. *ANSI Document, T1E1.4 Technical Subcommittee*, (98-173):1–12.

will improve if we reduce the energy of the distortion vector,

$$E\{|\mathbf{D}^{(\mathbf{X},g)} - \hat{\mathbf{D}}^{(\mathbf{X},g)}|^2\} < E\{|\mathbf{D}^{(\mathbf{X},g)}|^2\} \tag{6.31}$$

This property is demonstrated from the numerical examples in Section 3. for the simple uncoded case. If $\mathbf{D}^{(\mathbf{X},g)} \neq \hat{\mathbf{D}}^{(\mathbf{X},g)}$, we will not achieve perfect distortion cancellation in (6.27). However, if (6.31) is satisfied, the distortion will be smaller and the receiver will get a second improved estimate of $\mathbf{X}$ that it can use to reestimate $\mathbf{D}^{(\mathbf{X},g)}$.

The following describes a simple iterative algorithm based on hard decoding of the received vector but other alternatives are possible. For example, if the transceiver includes some error correction, the estimate of $\mathbf{X}$ will be more accurate and so will $\hat{\mathbf{D}}^{(\mathbf{X},g)}$. Performing hard decoding on the received vector $\mathbf{Y}^g$ will result in a simple estimate of the transmitted symbol $\mathbf{X}$:

$$\mathbf{X}^{(1)} = \left[\langle \frac{Y_0^g}{H_0} \rangle \cdots \langle \frac{Y_{N-1}^g}{H_{N-1}} \rangle \right]^T = \langle \mathbf{Y}^g / \mathbf{H} \rangle \tag{6.32}$$

where $\langle A_k \rangle$ denotes hard decoding, i.e. the operation of selecting the constellation point of tone k that is closest to A_k. From this estimate, we can compute, $\mathbf{x}^{(1)} = IDFT(\mathbf{X}^{(1)})$, and thus the estimate for $\mathbf{D}^{(\mathbf{X},g)}$ is:

$$\mathbf{D}^{(\mathbf{X}^{(1)},g)} = DFT(d_n^{(\mathbf{X}^{(1)},g)}) = DFT(g(x_n^{(1)}) - x_n^{(1)}) \tag{6.33}$$

If the SER is $p << 1$, a large fraction of the hard-decoded values will be correct since $Prob\{X_k^{(1)} = X_k\} = 1-p$. Therefore, $\mathbf{D}^{(\mathbf{X},g)} \approx \mathbf{D}^{(\mathbf{X}^{(1)},g)}$. Although the estimates may degrade from errors, the receiver can use these estimates to reduce the nonlinear distortion term in (6.23) in an iterative fashion. In this case, (6.27) becomes:

$$\mathbf{X}^{(q+1)} = \arg \min_{\bar{\mathbf{X}}} \|\mathbf{H} \circ \bar{\mathbf{X}} - \mathbf{H} \circ (\mathbf{X} + \mathbf{D}^{(\mathbf{X},g)} - \mathbf{D}^{(\mathbf{X}^{(q)},g)}) - \mathbf{N}\|^2 \tag{6.34}$$

or equivalently:

$$\mathbf{X}^{(q+1)} = \arg \min_{\bar{\mathbf{X}}} \|\bar{\mathbf{X}} - (\frac{\mathbf{Y}^g}{\mathbf{H}} - \mathbf{D}^{(\mathbf{X}^{(q)},g)})\|^2 \tag{6.35}$$

[Wulich, 1996] Wulich, D. (1996). Peak factor in orthogonal multicarrier modulation with variable levels. *Electronics Letters*, 32(20):1859–1860.

[Wulich et al., 1998] Wulich, D., Dinur, N., and Glinowiecki, A. (1998). Level clipped high order OFDM. Submitted for publication, 1998.

[Zekri and Van Biesen, 1999] Zekri, M. and Van Biesen, L. (1999). Super algorithm for clip probability reduction of DMT signals. In *Proceedings of the 13th International Conference on Information Networking(ICOIN-13)*, volume 2, pages 8B–3.1–6, Cheju Island, Korea.

[Zekri et al., 1998] Zekri, M., Van Biesen, L., and Spruyt, P. (1998). A new peak power reduction scheme for multicarrier modulation. In *Proc of IASTED Int Conf on Networks and Comm Syst*, Pittsburgh, PA.

[Zou and Wu, 1995] Zou, W. Y. and Wu, Y. (1995). COFDM: an overview. *IEEE Trans Broadcasting*, 41(1):1–8.

Index